Forschungsberichte aus dem Lehrstuhl für Regelungssysteme

Technische Universität Kaiserslautern

Band 10

Forschungsberichte aus dem Lehrstuhl

für Regelungssysteme

Technische Universität Kaiserslautern

Band 10

Herausgeber:

Prof. Dr. Steven Liu

Sven Reimann

Output-Based Control and Scheduling of Resource-Constrained Processes

Logos Verlag Berlin

Forschungsberichte aus dem Lehrstuhl für Regelungssysteme
Technische Universität Kaiserslautern

Herausgegeben von
Univ.-Prof. Dr.-Ing. Steven Liu
Lehrstuhl für Regelungssysteme
Technische Universität Kaiserslautern
Erwin-Schrödinger-Str. 12/332
D-67663 Kaiserslautern
E-Mail: sliu@eit.uni-kl.de

Bibliografische Information der Deutschen Nationalbibliothek

Die Deutsche Nationalbibliothek verzeichnet diese Publikation in der Deutschen Nationalbibliografie; detaillierte bibliografische Daten sind im Internet über http://dnb.d-nb.de abrufbar.

ISBN 978-3-8325-3980-1
ISSN 2190-7897

Logos Verlag Berlin GmbH
Comeniushof, Gubener Str. 47,
10243 Berlin
Tel.: +49 (0)30 / 42 85 10 90
Fax: +49 (0)30 / 42 85 10 92
http://www.logos-verlag.de

Output-Based Control and Scheduling of Resource-Constrained Processes

Ausgangsbasierte Regelungs- und Schedulingverfahren für Prozesse mit Ressourcenbeschränkungen

Vom Fachbereich Elektrotechnik und Informationstechnik
der Technischen Universität Kaiserslautern
zur Verleihung des akademischen Grades
Doktor der Ingenieurwissenschaften (Dr.-Ing.)
genehmigte Dissertation

von

Dipl.-Wirtsch.-Ing. Sven Reimann
geboren in Erlangen

D 386

Tag der mündlichen Prüfung:	23.03.2015
Dekan des Fachbereichs:	Prof. Dr.-Ing. Hans D. Schotten
Vorsitzender der Prüfungskommission:	Prof. Dr. techn. Gerhard Fohler
1. Berichterstatter:	Prof. Dr.-Ing. Steven Liu
2. Berichterstatter:	Prof. Dr.-Ing. Sandra Hirche

Acknowledgment

This thesis presents the results of my work at the Institute of Control Systems (LRS), Department of Electrical and Computer Engineering, at the University of Kaiserslautern.

Foremost, I would like to express my great gratitude to Prof. Dr.-Ing. Steven Liu, the head of the Institude of Control Systems, for the excellent supervision of my research, the scientific discussions, and also the good researching atmosphere.

Furthermore, I would like to thank Prof. Dr.-Ing. Sandra Hirche for her interest in my research and for joining the thesis committee as a reviewer. Thanks also go to Prof. Dr. techn. Gerhard Fohler for joining the thesis committee as a chair.

My time at the Institute of Control Systems has been very enjoyable and rewarding. I would like to thank all the colleagues for creating an open and cooperating atmosphere. Special thanks go to Priv. Doz. Dr.-Ing. habil. Christian Tuttas, Jun. Prof. Dr.-Ing. Daniel Görges, Dipl.-Ing. Fabian Kennel, Dipl.-Ing. Felix Berkel, M. Sc. Filipe Figueiredo, M. Sc. Hengyi Wang, Dr.-Ing. Jianfei Wang, Dr.-Ing. Liang Chen, M. Sc. Markus Bell, M. Sc. Markus Lepper, Dr.-Ing. Martin Pieschel, M. Sc. Michel Izák, Dipl.-Ing. Nadine Stegmann-Drüppel, Dipl.-Ing. Nelia Schneider, Dipl.-Ing. Peter Müller, Dr.-Ing. Philipp Münch, M. Sc. Sanad Al-Areqi, M. Eng. Sebastian Caba, Dr.-Ing. Stefan Simon, Dipl.-Ing. Tim Nagel, Dr.-Ing. Wei Wu, M. Sc. Xiaohai Lin, M. Sc. Yanhao He, and M. Sc. Yun Wan for a very good collaboration and many insightful scientific and personal discussions. Thanks also go to the technicians Swen Becker and Thomas Janz and to the secretary Jutta Lenhardt for providing a very good technical and administrative environment.

My biggest thanks go to my wife and my family for their support and love. This thesis is dedicated to them.

Kaiserslautern, April 2015 *Sven Reimann*

Contents

Notation

Throughout the thesis, scalars are denoted by lower- and upper-case non-bold letters $(a, b, \ldots, A, B, \ldots)$, vectors by lower-case bold letters $(\boldsymbol{a}, \boldsymbol{b}, \ldots)$, matrices by upper-case bold letters $(\boldsymbol{A}, \boldsymbol{B}, \ldots)$ and sets by upper-case double-struck letters $(\mathbb{A}, \mathbb{B}, \ldots)$.

Sets

$\mathbb{N}$	Set of positive integers
$\mathbb{N}_0$	Set of nonnegative integers
$\mathbb{R}$	Set of real numbers
$\mathbb{R}^+$	Set of positive real numbers
$\mathbb{R}_0^+$	Set of nonnegative real numbers
$\mathbb{R}^n$	Set of real vectors with dimension n
$\mathbb{R}^{n\times m}$	Set of real matrices with n rows and m columns

Operators

$\boldsymbol{A}^{-1}$	Inverse of matrix $\boldsymbol{A}$
$\boldsymbol{A}^T$	Transpose of matrix $\boldsymbol{A}$
$\boldsymbol{A}^{-T}$	Transpose of inverse of matrix $\boldsymbol{A}$, i.e. $(\boldsymbol{A}^{-1})^T = (\boldsymbol{A}^T)^{-1}$
$\boldsymbol{A} > 0$	Matrix $\boldsymbol{A} \in \mathbb{R}^{n\times n}$ positive definite, i.e. $\boldsymbol{x}^T\boldsymbol{A}\boldsymbol{x} > 0\ \forall \boldsymbol{x} \in \mathbb{R}^n \setminus \{\boldsymbol{0}\}$
$\boldsymbol{A} \geq 0$	Matrix $\boldsymbol{A} \in \mathbb{R}^{n\times n}$ positive semidefinite, i.e. $\boldsymbol{x}^T\boldsymbol{A}\boldsymbol{x} \geq 0\ \forall \boldsymbol{x} \in \mathbb{R}^n$
$\boldsymbol{A} < 0$	Matrix $\boldsymbol{A} \in \mathbb{R}^{n\times n}$ negative definite, i.e. $\boldsymbol{x}^T\boldsymbol{A}\boldsymbol{x} < 0\ \forall \boldsymbol{x} \in \mathbb{R}^n \setminus \{\boldsymbol{0}\}$
$\boldsymbol{A} \leq 0$	Matrix $\boldsymbol{A} \in \mathbb{R}^{n\times n}$ negative semidefinite, i.e. $\boldsymbol{x}^T\boldsymbol{A}\boldsymbol{x} \leq 0\ \forall \boldsymbol{x} \in \mathbb{R}^n$
$\mathrm{tr}(\boldsymbol{A})$	Trace of matrix $\boldsymbol{A}$
$\det(\boldsymbol{A})$	Determinant of matrix $\boldsymbol{A}$
$\lambda_{\min}(\boldsymbol{A})$	Minimum eigenvalue of matrix $\boldsymbol{A}$
$\lambda_{\max}(\boldsymbol{A})$	Maximum eigenvalue of matrix $\boldsymbol{A}$
$\mathrm{diag}(\boldsymbol{A}_1, \ldots)$	Block-diagonal matrix with blocks $\boldsymbol{A}_1, \ldots$
$\lVert\boldsymbol{x}\rVert_2$	Euclidean norm of vector $\boldsymbol{x} \in \mathbb{R}^n$, i.e. $\lVert\boldsymbol{x}\rVert_2 = \sqrt{\boldsymbol{x}^T\boldsymbol{x}} = \sqrt{x_1^2 + \ldots + x_n^2}$
$\mathrm{E}(X)$	Expected value of a random variable X
$\lfloor x \rfloor$	Floor, i.e. $\lfloor x \rfloor$ is the largest integer smaller than or equal to $x \in \mathbb{R}$
$x \bmod y$	Modulo, i.e. $x \bmod y = x - y \left\lfloor \frac{x}{y} \right\rfloor$ with $x, y \in \mathbb{R}$

Others

$\boldsymbol{I}$	Identity matrix
0	Zero matrix
$\left(\begin{smallmatrix} \boldsymbol{A} & * \\ \boldsymbol{B} & \boldsymbol{C} \end{smallmatrix}\right)$	Symmetric matrix $\left(\begin{smallmatrix} \boldsymbol{A} & \boldsymbol{B}^T \\ \boldsymbol{B} & \boldsymbol{C} \end{smallmatrix}\right)$

Acronyms

EDF	Earliest-Deadline First
ETC	Event-Triggered Control
EG	Event Generator
GAS	Globally Asymptotically Stable
GES	Globally Exponentially Stable
LQR	Linear-Quadratic Regulator
LMI	Linear Matrix Inequality
MAC	Medium Access Control
OF-CO-on	Output-Feedback Online Scheduling with Current Observer (Chapter 3)
OF-PO-on	Output-Feedback Online Scheduling with Prediction Observer (Chapter 3)
SF-off	State-Feedback Offline Scheduling (Chapter 3)
SF-on	State-Feedback Online Scheduling (Chapter 3)
ZOH	Zero-Order Hold

Linear Matrix Inequalities

Linear matrix inequalities (LMIs) are utilized throughout the thesis. Introductions to LMIs are given in [BEFB94] and [SP05] where also related topics like the S-procedure [BEFB94, Section 2.6.3], [SP05, Section 12.3.4], the Schur complement [BEFB94, pages 7-8], [SP05, Section 12.3.3] and the congruence transformation [SP05, Section 12.3.2] are addressed.

1 Introduction

1.1 Motivation

Modern control systems are usually realized as embedded systems to provide highly various functionality and good application performance. Such systems are also known as embedded control systems [ÅCH03]. An embedded system represents a computer in a physical environment, performing specific functions and reacting to the environment according to the requirements [ELLSV97]. Applications of embedded control systems can be found in various technical systems including transportation systems (e.g. automobiles and aircraft), consumer electronics (e.g. CD player), industrial applications (e.g. manufacturing and process control), and infrastructure (e.g. power systems and building automation) [Cer03, ÅC05, Gör12].

In embedded control systems, one processor typically handles several control and non-control tasks. Therefore, scheduling is required in order to decide which task to execute at a given time. Further, modern control systems are increasingly realized in a distributed manner, where sensors, actuators and controllers are located spatially distributed and communicate via a network. The application of such networked control systems is motivated by a larger flexibility and maintainability. On the other hand, networks require a medium access control (MAC) to schedule the communication resources.

Besides flexibility also resource constraints are a characteristic of embedded control systems. This has several reasons. First, embedded control systems are widely used in mass-market products. Due to the strong competition on the market, a particular attention during the development of embedded systems is put on the costs of the components. Those economic constraints then implicate for instance constraints on computing speed, memory size or communication bandwidth [ÅCH03]. Further, in networked control systems, the communication bandwidth is also limited based on physical constraints, which depend on the employed network [LMT01, SBR05]. For battery-powered mobile systems, with the nodes connected through a wireless network, also the energy consumed due to the transmissions plays an important role [SK97, FN01]. Therefore, the communication is limited to reduce the energy consumption. In this thesis, the focus lies on handling the control problem for systems with limited computation and communication resources. In order to distribute the limited resources efficiently, an intelligent scheduling algorithm is indispensable. Besides this, the control design plays an essential role. During the control design, the scheduling information needs to be taken into account to realize a good performance, which motivates an integrated control and scheduling design. In embedded control systems, the control and scheduling are typically realized in a time-triggered way. This means that the control update and scheduling are coordinated by a clock.

In recent years, event-triggered control has gained increasing attention in the context of networked control systems, as event-triggered control represents a promising strategy for handling the constraint resources, see e.g. [ÅB02, BHJ10, LL10, MT11, WL11]. Contrary to the well-established theory of digital control [ÅW90], the control input is not adjusted periodically but only when necessary. The necessity for an adjustment of the control input is specified by an event-triggering condition, which is purposely defined to reduce the communication load of the network, while maintaining a certain control performance. Challenges consist in the definition of a proper event-triggering condition and the control design suiting to the event-triggered implementation of the controller.

1.2 Objectives

The objective of this thesis consists in developing systematic approaches to using the limited computation and communication resources efficiently. Thereby, both scheduling approaches and event-triggered control approaches are studied as they both take the limitation of the resources explicitly into account. However, under such approaches the control input is adjusted aperiodically. This leads to the challenge of the design of suitable control parameters. Therefore, in this thesis a focus is put on the control synthesis considering explicitly the scheduling and event-triggered implementation. Besides the control performance, also the stability of the control systems needs to be ensured in the design. Priorities are set on the applicability of the methods with only output-feedback information and for PI controllers, which has not often been addressed. This is motivated by the facts that usually not all states are measurable in practical applications and that PI control is still one of the most widely used feedback control methods.

1.3 Related Work

After specifying the objectives, an overview of the related work is given. Thereby, the open points, which are covered in this thesis, are identified. First, scheduling algorithms for handling limited computation and communication resources are discussed. Second, approaches to event-triggered control are reviewed.

1.3.1 Control Task and Communication Network Scheduling

In practice the distribution of computation and communication resources is usually realized using real-time scheduling algorithms such as rate-monotonic (RM) or earliest-deadline first (EDF) [LL73] and medium access control protocols (MAC) protocols such as carrier sense multiple access (CSMA) or time division multiple access (TDMA). However, those scheduling algorithms are not designed specifically for control applications and have therefore drawbacks for control applications, such as time-varying sampling periods and latencies (jitter). These effects can essentially degrade the control performance or even lead to instability [CHL⁺03, CVHN09]. As a consequence of this, much research in recent years has focused on the control and scheduling design in a common framework [ÅC05, CA06, SSA10, LMV⁺13].

Those works can be classified into two categories, namely approaches to *sampling period assignment* and approaches to *integrated control and scheduling design*. The class of approaches to *sampling period assignment* focuses on handling limited computation resources. Control tasks are considered as periodic tasks, which have a fixed period, a known worst case execution time and a hard deadline. They are scheduled preemptively with standard scheduling algorithms, such as EDF. The idea of those approaches is to assign a sampling period to each control task such that the control performance is optimized. Therefore, an optimization problem is solved, where the solution is the sampling period. The sampling period assignment is either realized offline, before runtime, or online based on plant state information for instance. Those feedback scheduling approaches assume that the control parameters are given and they focus on limited computation resources and not on limited communication resources. On the other hand, under *integrated control and scheduling design* the control parameters are not assumed to be given but are designed jointly with the scheduler in a joint optimization problem. The solution of this optimization problem is a scheduling sequence and the control parameters of the control law. The scheduling is then realized non-preemptively with the determined sequence. Solutions are provided for distributing limited computation as well as communication resources. The benefit of the integrated control and scheduling design is that the controller and scheduler are harmonized with each other. This means that the jitter due to the scheduling is incorporated in the model and, hence, taken into account in the control design.

Approaches to Sampling Period Assignment

The publications [SLSS96, BC08, EHÅ00, CEBÅ02, MLB⁺04, MLB⁺09, HC05, CMV⁺06, CVMC11] focus on finding a tradeoff between sampling period and the quality of control subject to the resource constraints. Thereby, the control tasks are scheduled preemptively with an optimally assigned sampling period. In an early approach [SLSS96], it is assumed that the controllers are designed in the continuous-time domain and afterwards discretized for implementation. A performance index quantifying the performance of the digitalized control law of a given sampling period is introduced. The optimal sampling periods result from an offline optimization problem and are used for discretizing the controllers. This is extended in [BC08] by considering also the input delay besides the sampling period in the optimization problem. The input delay regards the delay from the sampling to the actuation, which is generally time-varying due to the preemptive scheduling.

Later publications extend the work of [SLSS96] to an online sampling period assignment [EHÅ00, CEBÅ02, MLB+04, MLB+09, HC05, CMV+06, CVMC11]. This is often referred to as feedback scheduling. In [EHÅ00], an online optimization method is proposed for adjusting the sampling period online based on the current processor load. Therefore, a state-feedback controller is considered and the LQ cost function is expressed as a function of the sampling period. This is extended in [CEBÅ02] to general linear dynamic controllers. Thus, the approaches proposed in [EHÅ00, CEBÅ02] focus on the resource allocation under a time-varying processor load.

In the approaches [MLB+04, MLB+09, HC05, CMV+06, CVMC11] the current plant states are incorporated in the online optimization, i.e. the sampling periods are adapted online based on a state-based performance index. The authors of [MLB+04, MLB+09] introduce a heuristic cost function taking current states into account for the sampling period assignment. Another approach in [HC05] makes use of a quadratic cost function related to the sampling period and the current state for adjusting the sampling period online. Henriksson *et al.* [HC05] assume a state-feedback control law whereas the approach in [CMV+06] generalizes it for linear dynamic controllers. Cervin *et al.* [CVMC11] complements the feedback scheduler from [CMV+06] by taking the computation delay into account and by adding a noise estimator as the evaluation of the cost function depends on the noise intensity. Further, a practical evaluation is given in [CVMC11].

In summary, the aim of those approaches is to select the sampling period such that a performance measure is minimized, while utilizing only a desired percentage of the processor resources. The performance measure is usually defined as a quadratic cost function. An open point for those approaches remains, how to design the control parameters. If the controller is designed based on the classical theory of digital control [ÅW90], the time-varying delays and sampling periods due to the preemption are not considered explicitly, which may lead to a poor control performance. Thus, the integrated control and scheduling design represents an alternative.

Approaches to Integrated Control and Scheduling Design

In a parallel line of research the integrated control and scheduling design is investigated in [RS00, LB02, GIL07, CA06, BÇH09] focusing on computation constraints and in [RS04, BÇH06, GIL09] focusing on communication constraints. The integrated design is also referred to as control and scheduling codesign. The approaches can be distinguished on the basis of offline and online scheduling, or also termed static and dynamic scheduling in literature. In the approaches [RS00, RS04, LB02, GIL07], an optimal state-feedback controller is designed jointly with a static sequence in an offline optimization problem. This means that the control tasks are scheduled non-preemptively according to the offline determined sequence, which is repeated perpetually, or in the networked case the control signals are transmitted and updated according to the sequence. In [RS00, RS04], a sequence with a defined periodicity is fixed, based on which the control design is formulated as a periodic linear quadratic control problem. The optimal sequence is then derived by exhaustive search among all sequence permutations. To make the problem computationally tractable a maximum periodicity is introduced. The work of [GIL07] extends this approach by considering the input delay of the control input, which is caused by the computation and transmission time. The method in [LB02] determines the optimal sequence and control via dynamic programming with tree pruning, to reduce combinatorial complexity. Offline scheduling has especially the advantage that it consumes few computation resources. However, it does not react to disturbances promptly.

Therefore, the approaches [CA06, BÇH06, BÇH09, GIL09] propose an extension to online scheduling, where the scheduling sequence is adapted online based on current plant state information. This allows to react to disturbances and thus, leads to an improved control performance. However, online scheduling may lead to a large overhead. Especially in case of limited computation resources, this may lead to conflicts, as the computation resources are scarce. In [CA06], the controller and scheduler are determined jointly by solving an infinite-horizon optimization problem using relaxed dynamic programming. However, this approach involves a large scheduling overhead. In [BÇH06, BÇH09], first an optimal periodic controller and a scheduling sequence are determined offline. Then, the starting index of the resulting scheduling sequence is adjusted online based on current plant state information [BÇH06]. The online adaption of the scheduling implicates considerable computational complexity, such that this algorithm is not applicable to limited computation resources. Therefore, in [BÇH09] a state-feedback scheduling mechanism is introduced, which adapts the scheduling sequence less frequently and in a suboptimal manner, such that the control performance is improved compared with the offline scheduling sequence with a bounded scheduling overhead. A receding-horizon control and scheduling codesign strategy is presented in [GIL09]. Thereby, the control and scheduling codesign problem is solved by optimizing a finite-horizon quadratic cost function using relaxed dynamic programming. The resulting scheduler and piecewise-linear state-feedback controller are applied online using the receding-horizon principle.

Similar to the approaches to sampling period assignment, the limited resources are distributed in a way such that a quadratic cost function measuring the control performance of all plants is minimized. Further, the aforecited codesign strategies are characterized by two properties. First, they assume that the plants are controlled using a state-feedback controller. Linear dynamic controllers, which are commonly applied in reality, are not considered. Second, the full state vector is required for the control and the online scheduling. This assumption of full state-feedback information is, however, not always realistic as the measurement of some states may not be possible or too noisy.

1.3.2 Event-Triggered Control

In embedded control systems, a plant is typically controlled by applying digital control theory, i.e. the control input is updated periodically with a fixed sampling interval. As the controller is implemented in a time-triggered way, the control input may be also updated even though there is no necessity, since the system is already in steady state for instance. This may lead to a deficient utilization of the limited computation and communication resources. Based on this observation, event-triggered control has been proposed in literature for an efficient utilization of the limited resources [Årz99, ÅB02]. The idea of event-triggered control is that control actions are only performed when necessary, e.g. when a certain evaluation condition of the controlled system is violated, rather than after the elapse of a fixed period. Therefore, an event generator (EG) is assigned to each sending node, where an event-triggering condition is verified. The communication network is then only utilized if an event is triggered.

Much research in recent years, therefore, has focused on event-triggered control design, see for instance [Tab07, LL10, WL11, MH09, Cog09, EDK10, HDT13, WRGL15, DH12, LL11a, HD13]. For the event-triggered control approaches in [Tab07, LL10, WL11, LL11a, DH12], the event-triggering condition needs to be monitored continuously. However, the continuous monitoring cannot be realized in standard time-sliced embedded software architectures. Further, the continuous-time event-triggered control requires the analysis of a minimum inter-event time to avoid the occurrence of Zeno-behavior. Zeno-behavior occurs, if consecutive events are triggered with an arbitrary small inter-event time, which is not realizable in practice. Therefore, discrete-time event-triggered control is introduced, where the event-triggering condition is not monitored continuously but periodically, see [MH09, Cog09, EDK10, HDT13, HD13, WRGL15]. Discrete-time event-triggered control can be easily implemented in standard time-sliced embedded software architectures and the non-occurrence of Zeno-behavior is directly guaranteed.

One of the fundamental challenges for the design of event-triggered control systems lies in the design of the control parameters which can offer satisfactory control performance while reducing the utilization of the limited resources. Often, an emulation-based design approach is followed. Emulation-based design is a two-step design procedure where the control parameters are first determined using classical digital control theory and afterwards the performance and stability of the closed-loop system is studied taking the event-triggered implementation of the controller into account. Approaches to a joint design of the controller and the event-triggering condition are proposed in [MH09, Cog09, WRGL15]. In [MH09, Cog09], an optimization problem with a cost function considering the expected quadratic control cost and the expected communication cost is established and solved using optimal stochastic control. These approaches start from a discrete-time plant model. In [WRGL15], a control synthesis minimizing a quadratic cost function is introduced based on a continuous-time plant model with input delay, taking the discrete-time event-triggered implementation into account

The results often consider state-feedback controllers assuming full state measurement, which are frequently not realizable. Exception can be found in [DH12, LL11a, HD13, Cog09], where in [DH12], an output-feedback event-triggered control scheme with a dynamic controller is introduced, and the state estimation is discussed in [LL11a, HD13, Cog09].

Event-Triggered PI Control

The aforecited event-triggered control strategies focus only on regulation problems, applying mostly static state-feedback controllers. Few researchers address the setpoint tracking problem in the event-triggered control framework, see [Årz99, OMT02, VK06, RJ08, DM09, SVD11, LL11b, TAJ12, LKJ12, KLJ14]. An early approach of event-triggered PI control [Årz99] shows that the communication effort can be significantly reduced with little degradation of the control performance compared with the time-based counterpart. As a structure of the event-triggering condition, an absolute threshold policy is proposed. The stability issue is, however, not addressed. Using absolute threshold policies may lead to serious problems, for instance sticking effect and large stationary oscillation as shown in [VK06, DM09]. To avoid the sticking effect, [VK06] proposes a sticking detection mechanism. A relatively complex absolute threshold policy taking also the integrator state into account is proposed in [TAJ12] to avoid both sticking and oscillation for stable first-order processes. For an asymptotic stability, an upper and lower bound on the inter-event time is considered. Taking the actuator saturation into account, stability analysis under absolute threshold policies is derived in [LKJ12, KLJ14], leading to a practical stability (ultimate boundedness). In addition to the setpoint tracking problem, the disturbance rejection problem is studied in [RJ08, SVD11, LL11b] under absolute threshold policies. In contrast, a relative threshold policy is proposed in [OMT02] with guaranteed exponential stability using perturbation-based stability analysis. Another relative threshold policy, guaranteeing asymptotic stability, is introduced in [GLS14], where the event-triggering condition is derived based on a linear quadratic performance criterion. Thus, a certain control performance can be ensured.

The aforecited event-triggered PI control strategies are characterized by considering an emulation-based design method. The design of the control parameters remains an open problem. Further, only few event-triggered PI control strategies propose discrete-time event-triggering conditions in the context of PI control [Årz99, VK06, DM09], whereas the benefits of a periodically sampled event-triggering scheme are well-known [HDT13]. Those approaches [Årz99, VK06, DM09] are characterized by two points, which will be advanced in this thesis. First, they use an absolute threshold policy which can bring challenges like sticking effect and oscillation. Second, a rigorous stability analysis is missing. First approaches towards the tuning of the PI control parameters under a specific event-triggered PI control scheme are introduced in [HPVK12, RJSD14]. However, the tuning rules are restricted to the control of first-order processes with a time delay.

In summary, many challenges occurring in the context of event-triggered PI control are discussed. However, they are usually only handled individually and not in common. Especially the design of the control parameters incorporating the event-triggered implementation of the PI controller remains a challenging point. Therefore, an event-triggered PI control scheme is proposed in this thesis, including a PI control synthesis based on a linear quadratic performance criterion. The proposed strategy allows to handle oscillation and sticking effects and requires only a periodic monitoring of the event-triggering condition. Finally, the input delay due to the transmission is also incorporated.

Regarding the evaluation of event-triggered control approaches, it needs to be considered that event-triggered control generally makes a trade-off between control performance and resource utilization. Therefore, the proposed methods include design parameters which allow to give more weight to the control performance or to the resource utilization. For evaluation, the control performance is to be contrasted with the resource utilization, see e.g. [Cog09, SVD11, TAJ12, AGL15]. The resource utilization can be rated by the number of triggered events [SVD11, TAJ12, WRGL15], which can also be converted into an average inter-event time as it is done in [AGL15]. For evaluating the control performance different measures are possible such as a quadratic cost function [Cog09, AGL15] or the integral of the absolute error [SVD11, TAJ12].

1.4 Outline and Contributions

The thesis consists of two parts. The first part (Chapter 2-4) focuses on control task and communication network scheduling, and the second part (Chapter 5) is devoted to event-triggered PI control. Each part can be read independently.

Chapter 2: In this chapter, the architecture of the control system with limited computation or communication resources is introduced. Thereby, the timing is discussed in detail, and the considered constraints are specified. This chapter makes preliminary formulations for Chapter 3 and Chapter 4.

Chapter 3: This chapter introduces a strategy to distribute the communication resources online based on output-feedback information. The networked control system is modeled as a discrete-time switched system. Based on the switched system model, a state-feedback controller and an online scheduler are first designed jointly in a codesign scheme, in order to minimize a linear quadratic cost function. At the second step, a prediction and a current observer are designed for estimating the states. As a switched system is handled, the validity of the separation principle needs to be verified. Further, it is shown that a certain performance can also be guaranteed if the control and scheduling are applied based on the estimated state vector instead of the real state vector. Besides an evaluation by simulation, where the proposed strategy is compared with an offline scheduling approach, also two experimental studies are presented to show the applicability and the benefits in a practical environment. Related concepts have been proposed in [CA06, BÇH09, GIL09], where, however, full state measurement is assumed for the control and online scheduling.

The presented strategy also makes a contribution in the context of discrete-time switched linear systems. For the design of the control and switching law of a discrete-time switched linear system, typically full state information is assumed, see e.g. [MFTM00, ZAHV09, GIL11]. The design of controller and observer for discrete-time switched linear systems is discussed in [DRI03, BMM09] applying the separation principle. However, the switching index is assumed arbitrary and does not follow a switching law. In [Zha01, GCB08], an output-based switching law is proposed for stabilizing an autonomous switched system. In the context of the stabilization of discrete-time switched systems, related concepts have been introduced in [JWX05, DGD11]. In [JWX05], a switched observer-based controller is designed for globally stabilizing the switched system. An $\mathcal{H}_\infty$ control approach is presented in [DGD11], where a switching rule and a switched dynamic output-feedback controller are designed jointly to stabilize the switched system. Additionally to those concepts, a linear quadratic performance measure is considered in the proposed strategy. This chapter is partly based on [RAL13b].

Chapter 4: This chapter proposes a novel online scheduling approach to PI control tasks based on output-feedback information for setpoint tracking problems. The aim is to design an online scheduler with a small scheduling overhead, such that it can efficiently distribute the limited computation resources. After introducing the non-preemptive online scheduling setup, the structure of the PI controller is presented. The PI controller is adapted automatically with respect to the time-varying control update interval, which is aroused by the scheduling. The scheduling strategy is then presented in Section 4.3. Subsequently, a systematic approach to designing the control parameters is proposed. Further, an a posteriori stability criterion is given for verifying stability. Finally, the effectiveness of the proposed method is illustrated by simulation and practical implementation. Comparisons are made with offline scheduling under a fixed sequence and with EDF scheduling. This chapter is partly based on [RWL14c], of which preliminary results appeared in [RWL14a].

Chapter 5: The focus of this chapter is the development of an event-triggered PI control scheme, which can handle several challenges of event-triggered PI control, simultaneously. After presenting the structure of the event-triggered control systems, two different event-triggering conditions are specified, which are analyzed afterwards. Those event-triggering conditions are characterized by a combined relative and absolute threshold, which is applied only in few publications, see e.g. [DH12]. Based on a linear quadratic cost function, a control synthesis is introduced for designing the control parameters, which takes the event-triggered implementation of the PI controller into account. Due to the additional absolute threshold, generally only practical stability can be achieved. Therefore, a practical stability criterion is given. The chapter is finalized by a detailed evaluation based on simulation and practical implementation. Comparisons are made with other event-triggered PI control concepts from literature. It is shown that oscillation and sticking can be handled by the proposed method and that an efficient implementation is possible due to the discrete-time monitoring of the event-triggering condition. The main contribution of this chapter is to develop an event-triggered PI control concept which includes a control synthesis. Previous publications on event-triggered control, including a control synthesis, mostly focus on event-triggered state-feedback control. This chapter is based on [RVAL15, RWL14b].

Chapter 6: The work of this thesis is summarized in this chapter. Additionally an outlook for possible future research directions is given.

1.5 Publications

Several publications on scheduling control tasks and communication networks as well as on event-triggered control have been finished during the doctoral studies. A chronological list of these articles together with their topic and their relation to this thesis is given below.

S. Reimann, W. Wu, and S. Liu. A novel control-schedule codesign method for embedded control systems. In *Proceedings of 2012 American Control Conference*, pages 3766–3771, 2012. (Control task/communication network scheduling)

S. Reimann, S. Al-Areqi, and S. Liu. An event-based online scheduling approach for networked embedded control systems. In *Proceedings of 2013 American Control Conference*, pages 5346–5351, 2013. (Communication network scheduling)

S. Al-Areqi, D. Görges, S. Reimann, and S. Liu. Event-based control and scheduling codesign of networked embedded control systems. In *Proceedings of 2013 American Control Conference*, pages 5319–5324, 2013. (Communication network scheduling and event-triggered control)

S. Reimann, S. Al-Areqi, and S. Liu. Output-based control and scheduling codesign for control systems sharing a limited resource. In *Proceedings of the 52nd IEEE Conference on Decision and Control*, pages 4042–4047, 2013. (Control task/communication network scheduling, Chapter 3)

S. Reimann, W. Wu, and S. Liu. PI control and scheduling design for embedded control systems. In *Proceedings of 19th IFAC World Congress*, pages 11111–11116, 2014. (Control task scheduling, Chapter 4)

W. Wu, S. Reimann, and S. Liu. Event-triggered control for linear systems subject to actuator saturation. In *Proceedings of 19th IFAC World Congress*, pages 9492–9497, 2014. (Event-triggered control)

W. Wu, S. Reimann, D. Görges, and S. Liu. Suboptimal event-triggered control for time-delayed linear systems. *IEEE Transactions on Automatic Control*, to appear, 2015. (Event-triggered control)

S. Reimann, D. H. Van, S. Al-Areqi, and S. Liu. Stability Analysis and PI Control

Synthesis under Event-Triggered Communication. In *Proceedings of the 2015 European Control Conference*, 2015. (Event-triggered control, Chapter 5)

B. Watkins, S. Al-Areqi, S. Reimann, and S. Liu. Event-Based Control of Constrained Discrete-Time Linear Systems with Guaranteed Performance. *International Journal of Sensors, Wireless Communications and Control*, to appear, 2015. (Event-triggered control)

Part I

Control Task and Communication Network Scheduling

2 Control Task and Communication Network Scheduling Model

In this chapter, the two control and scheduling problems with resource constraints are introduced. The first problem (Section 2.1) handles limited computation resources, whereas in the second problem (Section 2.2) limited communication resources are dealt with.

2.1 Control Task Scheduling

Consider a set of M independent linear time-invariant plants P_i, $i \in \{1, ..., M\}$, controlled by a set of control tasks T_i, $i \in \{1, ..., M\}$, using a single embedded processor. For distributing the limited resources a scheduler S is required. The general structure is given in Figure 2.1. The aim is to design a non-preemptive scheduler and a controller for each control task based on output-feedback information.

In the following the different steps for the realization of the scheduler and the controller

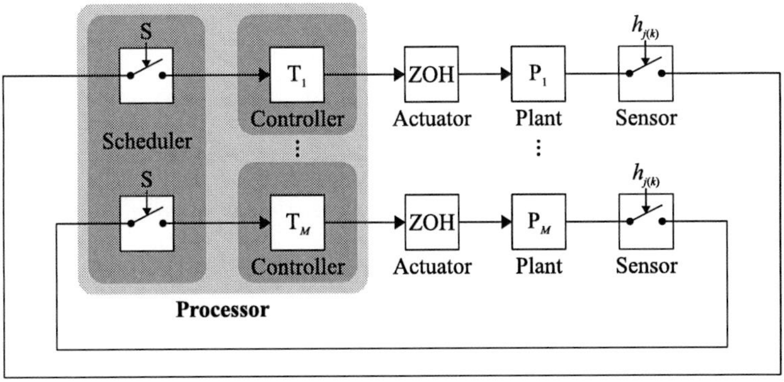

Figure 2.1: Control and scheduling architecture of a control system with computation constraints

are discussed. Thereby, six steps are distinguished:

1. Sampling the output vector $\boldsymbol{y}_i(t_k)$ or the measurable state vector $\boldsymbol{x}_{\mathrm{p}i}(t_k)$, respectively, of each plant P_i at the time instant t_k and transmitting them to the processor. (Execution time τ_{SC})

2. Correcting the prediction for the estimation of the state vector $\boldsymbol{x}_{\mathrm{p}i}(t_k)$ of each plant P_i in case of a current observer. This step is not required if all states are measurable or the scheduling decision is made based on the output vector. (Execution time $\tau_{\mathrm{O/cor}}$)

3. Determining the control task index $j(k) \in \mathbb{J} = \{1, ..., M\}$, i.e. the index of the plant for which the control task is executed. (Execution time τ_{S})

4. Computing the control input $\boldsymbol{u}_{j(k)}(t_k)$ of the plant $\mathrm{P}_{j(k)}$. (Execution time $\tau_{\mathrm{C}j(k)}$)

5. Applying the computed control input $\boldsymbol{u}_{j(k)}(t_k)$ to the plant $\mathrm{P}_{j(k)}$. For the other plants $\mathrm{P}_i \neq \mathrm{P}_{j(k)}$ the control signal is held constant. (Execution time $\tau_{\mathrm{CA}j(k)}$)

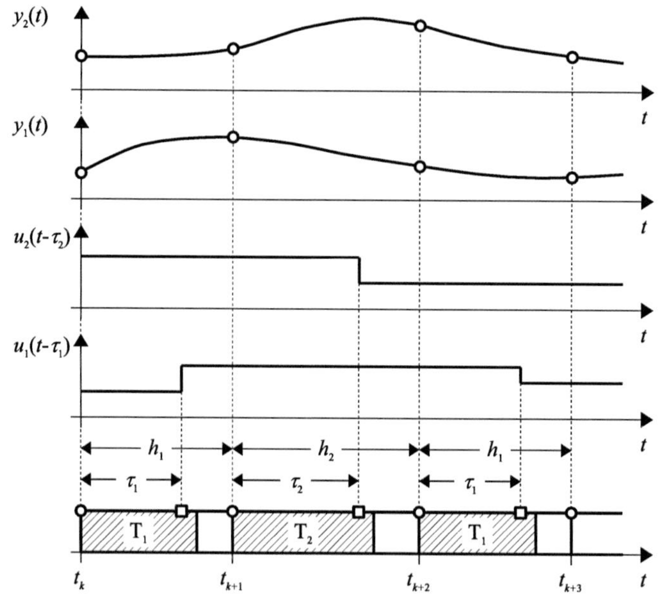

Figure 2.2: Scheduling timing diagram for two plants where a new measurement is indicated by a circle, a control update by a square, and the execution of a control task T_i by the hatching

6. Predicting the estimated state vector of each plant P_i. This step is not required if all states are measurable or the scheduling decision is made based on the output vector. (Execution time $\tau_{O/pre}$)

The execution time of each step is relevant for determining the input delay τ_i, i.e. the time delay from the measurement to the actuation, which is considered in the model (3.1), and for analyzing the resource utilization, which will be discussed in Section 4.6.2. Thus, the input delay is given by

$$\tau_i = \tau_{SC} + \tau_{O/cor} + \tau_S + \tau_{Ci} + \tau_{CAi}. \tag{2.1}$$

As the execution time of each step is quite static and predictable the input delay τ_i can be assumed constant.

The following measurement of the output is realized at the time instant $t_{k+1} = t_k + h_{j(k)}$ where $h_{j(k)}$ is the step size of the executed task $T_{j(k)}$. The step size $h_{j(k)}$ is chosen larger than or equal to the overall execution time $\tau_{j(k)} + \tau_{O/pre}$, i.e. $h_{j(k)} \geq \tau_{j(k)} + \tau_{O/pre}$. Since the input delay can be different for each control task, the step size can be selected time-varying with respect to $j(k)$. The remaining idle time can be devoted to executing non-control tasks. The resulting timing is illustrated in Figure 2.2.

The control task scheduling problem is solved in Chapter 4 for PI control tasks.

2.2 Communication Network Scheduling

In the second problem the architecture in Figure 2.1 is extended by including a single channel communication network, i.e. the control signal is sent from the processor to

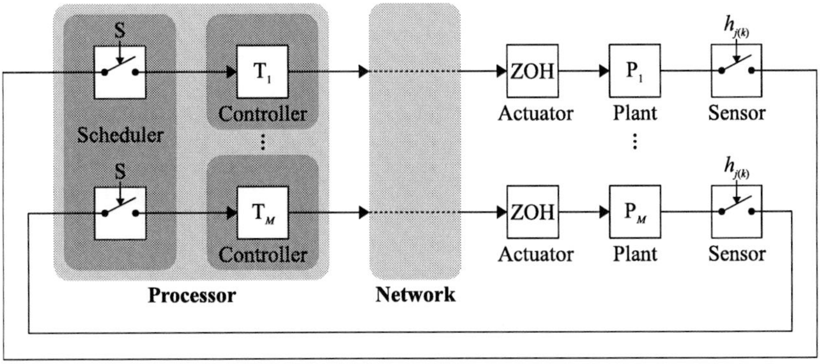

Figure 2.3: Control and scheduling architecture of a control system with communication constraints

the distant actuator over a limited bandwidth communication network, as shown in Figure 2.3. The focus lies on the information exchange between the controller and the actuator. Thus, the scheduler decides for which plant the control signal is computed and transmitted to the corresponding actuator. The discussed realization steps in Section 2.1 are likewise valid for this problem. However in Step 5, the control signal is transmitted over a network, which may lead to a larger delay $\tau_{\mathrm{CA}i}$. It is assumed that the variation of the transmission time is negligible. Packet loss and quantization effects are also not considered.

The approach in Chapter 3 focuses on scheduling online the controller-actuator link, which requires full output information of all plants. In real implementations, also the sensor information may be transmitted over a communication network to the controller such that the sensor-controller link needs to be scheduled. However, the online scheduling problem of the sensor-controller link differs essentially from the online scheduling problem of the controller-actuator link and also depends on the communication network. Different approaches are required for this problem, which is not subject to this thesis. Further discussion on scheduling the sensor-controller link is given in [Ben06, Section 4.1] and [Gör12, Section 11.2].

3 Output-Feedback Control and Scheduling Design

In this chapter, first a state-based non-preemptive scheduler and a state-feedback controller are designed jointly in a codesign procedure. This means that based on full state information of all plants the scheduler decides which control signal is transmitted over the communication network. In order to make a scheduling decision based on the output vector, an observer is included for estimating the states. As this approach requires a large scheduling overhead it is not suitable for control systems subject to limited computation resources (Section 2.1) but for control systems subject to limited communication resources (Section 2.2).

3.1 Modeling

3.1.1 Continuous-Time Model

Each plant P_i, $i \in \mathbb{J} = \{1, ..., M\}$, is described by the continuous-time state equation

$$\begin{aligned} \dot{\boldsymbol{x}}_{\mathrm{p}i}(t) &= \boldsymbol{A}_{\mathrm{p}i}\boldsymbol{x}_{\mathrm{p}i}(t) + \boldsymbol{B}_{\mathrm{p}i}\boldsymbol{u}_i(t-\tau_i) \\ \boldsymbol{y}_i(t) &= \boldsymbol{C}_{\mathrm{p}i}\boldsymbol{x}_{\mathrm{p}i}(t) \end{aligned} \tag{3.1}$$

where $\boldsymbol{A}_{\mathrm{p}i} \in \mathbb{R}^{n_i \times n_i}$ is the system matrix, $\boldsymbol{B}_{\mathrm{p}i} \in \mathbb{R}^{n_i \times m_i}$ is the input matrix, $\boldsymbol{C}_{\mathrm{p}i} \in \mathbb{R}^{p_i \times n_i}$ is the output matrix, $\boldsymbol{x}_{\mathrm{p}i}(t) \in \mathbb{R}^{n_i}$ is the state vector, $\boldsymbol{y}_i(t) \in \mathbb{R}^{p_i}$ is the output vector and $\boldsymbol{u}_i(t-\tau_i) \in \mathbb{R}^{m_i}$ is control vector with the constant input delay τ_i defined in (2.1).

A continuous-time cost function associated to each plant P_i is given by

$$J_i = \int_0^\infty \begin{pmatrix} \boldsymbol{x}_{\mathrm{p}i}(t) \\ \boldsymbol{u}_i(t-\tau_i) \end{pmatrix}^T \begin{pmatrix} \boldsymbol{Q}_{\mathrm{c}i} & \boldsymbol{0} \\ \boldsymbol{0} & \boldsymbol{R}_{\mathrm{c}i} \end{pmatrix} \begin{pmatrix} \boldsymbol{x}_{\mathrm{p}i}(t) \\ \boldsymbol{u}_i(t-\tau_i) \end{pmatrix} dt \tag{3.2}$$

where $\boldsymbol{Q}_{ci} \in \mathbb{R}^{n_i \times n_i}$ is symmetric and positive semidefinite and $\boldsymbol{R}_{\mathrm{c}i} \in \mathbb{R}^{m_i \times m_i}$ is symmetric and positive definite. The objective is to design a scheduler and a controller jointly, such that a certain control performance can be guaranteed. As performance measure, the overall cost function $J = \sum_{i=1}^{M} J_i$ is introduced.

3.1.2 Discrete-Time Model

Due to the discrete nature of the computation platform, a discrete-time representation of the continuous-time state equation (3.1) and cost function (3.2) is crucial. The discretization is realized over the discretization interval $t_k \leq t < t_{k+1}$ using ZOH. The discretization interval is given by the step size $h_{j(k)} = t_{k+1} - t_k$ of the executed task $\mathrm{T}_{j(k)}$ which is chosen larger than or equal to the corresponding input delay $\tau_{j(k)}$, as discussed in Chapter 2.

For the discretization, it must be distinguished whether the control task T_i of the considered plant P_i is executed or not within the discretization interval $t_k \leq t < t_{k+1}$. If the control task T_i of the considered plant P_i is executed ($i = j(k)$), then the control signal is updated within $t_k \leq t < t_{k+1}$, i.e.

$$\boldsymbol{u}_i(t-\tau_i) = \begin{cases} \boldsymbol{u}_i(t_{k-1}) & \text{for} \quad t_k \leq t < t_k + \tau_i \\ \boldsymbol{u}_i(t_k) & \text{for} \quad t_k + \tau_i \leq t < t_{k+1}. \end{cases} \tag{3.3}$$

If the control task T_i of the considered plant P_i is not executed ($i \neq j(k)$), then the control signal is not updated at all and held constant, i.e.

$$\boldsymbol{u}_i(t-\tau_i) = \boldsymbol{u}_i(t_{k-1}) \quad \text{for} \quad t_k \leq t < t_{k+1}. \tag{3.4}$$

The logical variable

$$\delta_i(k) = \begin{cases} 1 & \text{if} \quad i = j(k) \\ 0 & \text{if} \quad i \neq j(k) \end{cases} \tag{3.5}$$

is introduced to indicate if the control task T_i of the considered plant P_i is executed or not. Defining the generalized delay

$$\check{\tau}_{ij(k)} = \delta_i(k)\tau_i + (1-\delta_i(k))h_{j(k)} \tag{3.6}$$

allows to formulate a discrete-time state equation corresponding to (3.1) with respect to the task index $j(k)$ based on an augmented state vector as

$$\begin{aligned} \boldsymbol{x}_{\mathrm{a}i}(k+1) &= \boldsymbol{A}_{\mathrm{a}ij(k)}\boldsymbol{x}_{\mathrm{a}i}(k) + \boldsymbol{B}_{\mathrm{a}ij(k)}\boldsymbol{u}_i(k) \\ \boldsymbol{y}_i(k) &= \boldsymbol{C}_{\mathrm{a}i}\boldsymbol{x}_{\mathrm{a}i}(k) \end{aligned} \tag{3.7}$$

where the state vector $\boldsymbol{x}_{\mathrm{a}i}(k) \in \mathbb{R}^{n_i+m_i}$, the system matrix $\boldsymbol{A}_{\mathrm{a}ij(k)} \in \mathbb{R}^{(n_i+m_i)\times(n_i+m_i)}$, the input matrix $\boldsymbol{B}_{\mathrm{a}ij(k)} \in \mathbb{R}^{(n_i+m_i)\times m_i}$ and the output matrix $\boldsymbol{C}_{\mathrm{a}i} \in \mathbb{R}^{p_i\times(n_i+m_i)}$ are defined as

$$\boldsymbol{x}_{\mathrm{a}i}(k) = \begin{pmatrix} \boldsymbol{x}_{\mathrm{p}i}(k) \\ \boldsymbol{u}_i(k-1) \end{pmatrix} \tag{3.8a}$$

$$\boldsymbol{A}_{\mathrm{a}ij(k)} = \begin{pmatrix} \boldsymbol{\Phi}_{ij(k)} & \boldsymbol{\Gamma}_{1ij(k)} \\ 0 & (1-\delta_i(k))\boldsymbol{I} \end{pmatrix} \tag{3.8b}$$

$$\boldsymbol{B}_{\mathrm{a}ij(k)} = \begin{pmatrix} \boldsymbol{\Gamma}_{0ij(k)} \\ \delta_i(k)\boldsymbol{I} \end{pmatrix} \tag{3.8c}$$

$$\boldsymbol{C}_{\mathrm{a}i} = \begin{pmatrix} \boldsymbol{C}_{\mathrm{p}i} & 0 \end{pmatrix} \tag{3.8d}$$

with

$$\Phi_{ij(k)} = e^{A_{\mathrm{p}i} h_{j(k)}} \tag{3.9a}$$

$$\Gamma_{1ij(k)} = \int_{h_{j(k)} - \check{\tau}_{ij(k)}}^{h_{j(k)}} e^{A_{\mathrm{p}i} s} ds B_{\mathrm{p}i} \tag{3.9b}$$

$$\Gamma_{0ij(k)} = \int_{0}^{h_{j(k)} - \check{\tau}_{ij(k)}} e^{A_{\mathrm{p}i} s} ds B_{\mathrm{p}i}. \tag{3.9c}$$

This representation of systems with delayed inputs is adapted from [ÅW90, Section 3.2].

The overall system can then be modeled with respect to the task index $j(k)$ by the block-diagonal discrete-time switched linear system

$$\begin{aligned} x(k+1) &= A_{j(k)} x(k) + B_{j(k)} u(k) \\ y(k) &= C x(k) \end{aligned} \tag{3.10}$$

where the system matrix $A_{j(k)} \in \mathbb{R}^{(n+m)\times(n+m)}$, the input matrix $B_{j(k)} \in \mathbb{R}^{(n+m)\times m}$, the output matrix $C \in \mathbb{R}^{p\times(n+m)}$, the state vector $x(k) \in \mathbb{R}^{n+m}$, the input vector $u(k) \in \mathbb{R}^{m}$, and the output vector $y(k) \in \mathbb{R}^{p}$ with

$$n = \sum_{i=1}^{M} n_i, \quad m = \sum_{i=1}^{M} m_i, \quad p = \sum_{i=1}^{M} p_i \tag{3.11}$$

are given by

$$x(k) = \left(x_{\mathrm{a}1}^T(k) \quad \cdots \quad x_{\mathrm{a}M}^T(k) \right)^T \tag{3.12a}$$

$$u(k) = \left(u_1^T(k) \quad \cdots \quad u_M^T(k) \right)^T \tag{3.12b}$$

$$y(k) = \left(y_1^T(k) \quad \cdots \quad y_M^T(k) \right)^T \tag{3.12c}$$

$$A_{j(k)} = \mathrm{diag}\left(A_{\mathrm{a}1j(k)}, \ldots, A_{\mathrm{a}Mj(k)} \right) \tag{3.12d}$$

$$B_{j(k)} = \mathrm{diag}\left(B_{\mathrm{a}1j(k)}, \ldots, B_{\mathrm{a}Mj(k)} \right) \tag{3.12e}$$

$$C = \mathrm{diag}\left(C_{\mathrm{a}1}, \ldots, C_{\mathrm{a}M} \right). \tag{3.12f}$$

The continuous-time cost function (3.2) is discretized in an analogous manner, i.e. it is distinguished whether the control task T_i of the considered plant P_i is executed or not within $t_k \leq t < t_{k+1}$. This leads to the discrete-time cost function

$$J_i = \sum_{k=0}^{\infty} \begin{pmatrix} x_{\mathrm{p}i}(k) \\ u_i(k-1) \\ u_i(k) \end{pmatrix}^T Q_{ij(k)} \begin{pmatrix} x_{\mathrm{p}i}(k) \\ u_i(k-1) \\ u_i(k) \end{pmatrix} \tag{3.13}$$

associated to the plant P_i with the weighting matrix $Q_{ij(k)} \in \mathbb{R}^{(n_i+2m_i)\times(n_i+2m_i)}$ defined in equation (A.2).

The overall cost function, defined as the sum of the individual cost functions, i.e.

$$J = \sum_{i=1}^{M} J_i, \tag{3.14}$$

can then be written as

$$J = \sum_{k=0}^{\infty} \begin{pmatrix} \boldsymbol{x}(k) \\ \boldsymbol{u}(k) \end{pmatrix}^T \boldsymbol{Q}_{j(k)} \begin{pmatrix} \boldsymbol{x}(k) \\ \boldsymbol{u}(k) \end{pmatrix}. \tag{3.15}$$

The weighting matrix $\boldsymbol{Q}_{j(k)} \in \mathbb{R}^{(n+2m)\times(n+2m)}$, which is symmetric and positive semidefinite, is computed according to equation (A.4). More details on the discretization procedure of the cost function (3.2) are given in the Appendix A.1.1.

3.2 Problem Formulation

The objective is to find a control and scheduling law which can be realized based on output feedback. In the following, this is achieved in a two-step design procedure applying the separation principle. In the first step a control and scheduling law is determined jointly assuming state-feedback information. This codesign problem is formulated as

Problem 3.1 *For the discrete-time switched system* (3.10) *find a state-feedback control law* $\boldsymbol{u}(k)$ *and a scheduling law* $j(k)$ *such that the closed-loop system is globally asymptotically stable (GAS) with guaranteed performance.*

In the second step a Luenberger observer is designed such the control and scheduling law can be realized based on the estimated state vector $\bar{\boldsymbol{x}}(k)$. The observer design problem is formulated as

Problem 3.2 *For the discrete-time switched system* (3.10) *find a Luenberger observer for estimating the states such that the observation error is exponentially stabilized with a decay rate* ρ.

The resulting output-based control and scheduling architecture is shown in Figure 3.1.

As the control and scheduling law is designed assuming the true state vector $\boldsymbol{x}(k)$ is fed back, it needs to be verified if the classical separation principle [FPW98, Section 8.3.1] still holds in the proposed framework, which is discussed in Section 3.5.

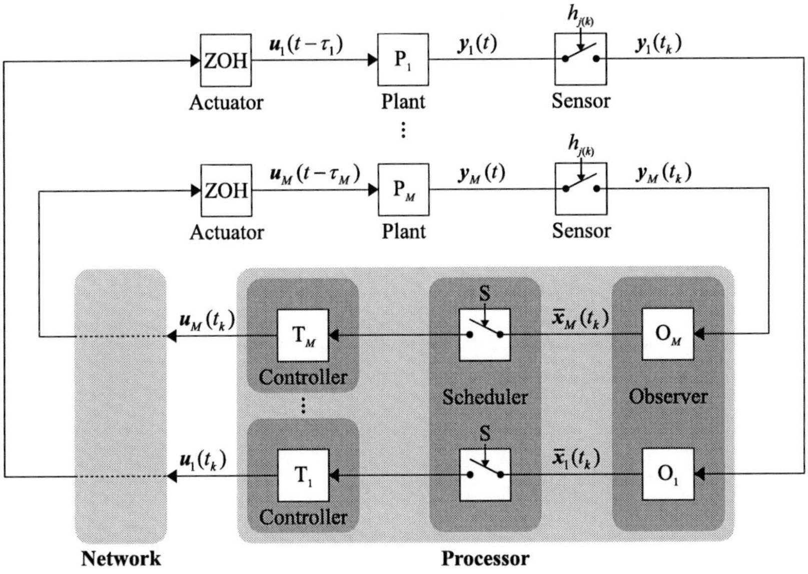

Figure 3.1: Output-based control and scheduling architecture

3.3 State-Feedback Control and Scheduling Codesign

Consider the state-feedback control law

$$\boldsymbol{u}(k) = \boldsymbol{K}_{j(k)}\boldsymbol{x}(k) \tag{3.16}$$

with the switched control gain matrix $\boldsymbol{K}_{j(k)} \in \mathbb{R}^{m\times(n+m)}$. Further, consider the scheduler with the state-feedback scheduling law

$$j(k) = \arg\min_{l\in\mathbb{J}} \boldsymbol{x}^T(k)\boldsymbol{P}_l\boldsymbol{x}(k) \tag{3.17}$$

with $\boldsymbol{P}_l \in \mathbb{R}^{(n+m)\times(n+m)}$ symmetric and positive definite. The following derivations show that this scheduling law is crucial to stabilize the switched system (3.10).

Problem 3.3 *For the discrete-time switched system* (3.10), *find the control gain matrices* $\boldsymbol{K}_i$ *of the control law* (3.16) *and the scheduling matrices* $\boldsymbol{P}_i$ *of the scheduling law* (3.17) *with* $i \in \mathbb{J}$ *such that the closed-loop system is globally asymptotically stable (GAS) with guaranteed performance.*

Remark 3.1. Due to the defined zero-order hold property of the preceding control vector $\boldsymbol{u}_i(k-1)$ in case a control task T_i is not executed ($\delta_i(k) = 0$), see equation (3.4), the system matrix $\boldsymbol{A}_{\mathrm{a}ij(k)}$ has eigenvalues equal 1 for $\delta_i(k) = 0$. Additionally, the input matrix is $\boldsymbol{B}_{\mathrm{a}ij(k)} = \boldsymbol{0}$ and thus, the pair $(\boldsymbol{A}_{\mathrm{a}ij(k)}, \boldsymbol{B}_{\mathrm{a}ij(k)})$ is not stabilizable for $\delta_i(k) = 0$. Because of the block-diagonal structure of the system and input matrix of the switched system (3.10) and the therefore decoupled structure, this means that the subsystem of the switched system (3.10) defined by the pair $(\boldsymbol{A}_{j(k)}, \boldsymbol{B}_{j(k)})$ is not stabilizable for any $j(k) \in \mathbb{J}$. Consequently, the closed-loop system matrix $\boldsymbol{A}_{j(k)} + \boldsymbol{B}_{j(k)}\boldsymbol{K}_{j(k)}$ is not stable for any $j(k) \in \mathbb{J}$. In order to stabilize a switched system consisting of possibly unstable subsystems based on a state-based switching law, different approaches have been proposed in the context of discrete-time systems, e.g. [Zha01, GC06, LA06]. The scheduling law (3.17) is motivated by the stabilizing switching law proposed in [GC06]. A different control and scheduling codesign procedure motivated by the results of [Zha01] is presented in [RWL12].

Substituting the control law (3.16) into (3.10) leads to the discrete-time closed-loop switched system

$$\boldsymbol{x}(k+1) = \left(\boldsymbol{A}_{j(k)} + \boldsymbol{B}_{j(k)}\boldsymbol{K}_{j(k)}\right)\boldsymbol{x}(k) = \tilde{\boldsymbol{A}}_{j(k)}\boldsymbol{x}(k). \tag{3.18}$$

Substituting further (3.16) into (3.15) results in the closed-loop cost function

$$J = \sum_{k=0}^{\infty} \boldsymbol{x}^T(k) \begin{pmatrix} \boldsymbol{I} \\ \boldsymbol{K}_{j(k)} \end{pmatrix}^T \boldsymbol{Q}_{j(k)} \begin{pmatrix} \boldsymbol{I} \\ \boldsymbol{K}_{j(k)} \end{pmatrix} \boldsymbol{x}(k) = \sum_{k=0}^{\infty} \boldsymbol{x}^T(k)\tilde{\boldsymbol{Q}}_{j(k)}\boldsymbol{x}(k). \tag{3.19}$$

Motivated by the scheduling law (3.17) the piecewise quadratic Lyapunov function

$$V_{\mathrm{c}}(k) = \min_{l \in \mathbb{J}} \boldsymbol{x}^T(k)\boldsymbol{P}_l\boldsymbol{x}(k) \tag{3.20}$$

with $\boldsymbol{P}_l \in \mathbb{R}^{(n+m)\times(n+m)}$ symmetric and positive definite is introduced, which is positive definite and radially unbounded since

$$\alpha_1 \left\|\boldsymbol{x}(k)\right\|_2^2 \leq V_{\mathrm{c}}(k) \leq \alpha_2 \left\|\boldsymbol{x}(k)\right\|_2^2 \tag{3.21}$$

with

$$\alpha_1 = \min_{l \in \mathbb{J}} \lambda_{\min}(\boldsymbol{P}_l), \quad \alpha_2 = \max_{l \in \mathbb{J}} \lambda_{\max}(\boldsymbol{P}_l) \tag{3.22}$$

holds for all $\boldsymbol{x} \in \mathbb{R}^{n+m}$. Under given control gain matrices $\boldsymbol{K}_i$, $i \in \mathbb{J}$, the stability of the closed-loop switched system can be analyzed based on the following theorem, see also [GC06].

Theorem 3.1 *The discrete-time closed-loop switched system* (3.18) *is GAS under the scheduling law* (3.17) *if there exist a set of matrices $\boldsymbol{P}_q \in \mathbb{R}^{(n+m)\times(n+m)}$ symmetric and positive definite and a set of scalars $\mu_{lq} \in \mathbb{R}_0^+$, with $q, l \in \mathbb{J}$, satisfying the matrix inequality*

$$\tilde{\boldsymbol{A}}_q^T \left(\sum_{l=1}^{M} \mu_{lq}\boldsymbol{P}_l\right) \tilde{\boldsymbol{A}}_q - \boldsymbol{P}_q < -\tilde{\boldsymbol{Q}}_q \tag{3.23}$$

and the equality $\sum_{l=1}^{M} \mu_{lq} = 1$ *for all* $q \in \mathbb{J}$. *Furthermore, the closed-loop cost function* (3.19) *is bounded.*

PROOF. Assume that the matrix inequality (3.23) is feasible for all $q \in \mathbb{J}$. Since $\tilde{\boldsymbol{Q}}_q$ is symmetric and positive semidefinite the inequality

$$\tilde{\boldsymbol{A}}_q^T \left(\sum_{l=1}^{M} \mu_{lq} \boldsymbol{P}_l \right) \tilde{\boldsymbol{A}}_q - \boldsymbol{P}_q < 0 \tag{3.24}$$

also holds. Assuming the resulting scheduling index given by (3.17) at a time instant t_k is $j(k) = q$ for some $q \in \mathbb{J}$, the Lyapunov function (3.20) at the time instant t_{k+1} is calculated as

$$\begin{aligned} V_c(k+1) &= \min_{l \in \mathbb{J}} \boldsymbol{x}^T(k) \tilde{\boldsymbol{A}}_q^T \boldsymbol{P}_l \tilde{\boldsymbol{A}}_q \boldsymbol{x}(k) \\ &= \min_{\mu_{1q},\dots\mu_{Mq}} \boldsymbol{x}^T(k) \tilde{\boldsymbol{A}}_q^T \left(\sum_{l=1}^{M} \mu_{lq} \boldsymbol{P}_l \right) \tilde{\boldsymbol{A}}_q \boldsymbol{x}(k) \\ &\leq \boldsymbol{x}^T(k) \tilde{\boldsymbol{A}}_q^T \left(\sum_{l=1}^{M} \mu_{lq} \boldsymbol{P}_l \right) \tilde{\boldsymbol{A}}_q \boldsymbol{x}(k). \end{aligned} \tag{3.25}$$

Combining condition (3.24) and (3.25) it can be concluded that

$$V_c(k+1) < \boldsymbol{x}^T(k) \boldsymbol{P}_q \boldsymbol{x}(k) = V_c(k) \tag{3.26}$$

holds for all $\boldsymbol{x}(k) \neq \boldsymbol{0}$, which implies that the closed-loop system (3.18) is GAS.

Pre-/post-multiplying (3.23) by the vector $\boldsymbol{x}^T(k)$ and its transpose, respectively, and considering condition (3.25) results in

$$\min_{l \in \mathbb{J}} \boldsymbol{x}^T(k) \left(\tilde{\boldsymbol{A}}_{j(k)}^T \boldsymbol{P}_l \tilde{\boldsymbol{A}}_{j(k)} - \boldsymbol{P}_{j(k)} \right) \boldsymbol{x}(k) < -\boldsymbol{x}^T(k) \tilde{\boldsymbol{Q}}_{j(k)} \boldsymbol{x}(k). \tag{3.27}$$

Summing up (3.27) over $k = 0, ..., \infty$ yields

$$V_c(0) - \lim_{k \to \infty} V_c(k) > J. \tag{3.28}$$

As the closed-loop system (3.18) is GAS, $V_c(k) \to 0$ for $k \to \infty$. Thus, inequality (3.28) implies an upper bound

$$J < \min_{l \in \mathbb{J}} \boldsymbol{x}^T(0) \boldsymbol{P}_l \boldsymbol{x}(0) < \mathrm{tr}(\boldsymbol{P}_{j(0)}) \, \|\boldsymbol{x}(0)\|_2^2 . \tag{3.29}$$

This completes the proof. □

Based on Theorem (3.1) an LMI optimization problem is derived to design jointly the control and scheduling parameters $\boldsymbol{K}_q$ and $\boldsymbol{P}_q$ while minimizing the upper bound (3.29).

Theorem 3.2 *The solution to Problem 3.3 is obtained from the LMI optimization problem*

$$\min_{\boldsymbol{P}_q, \boldsymbol{K}_q} \operatorname{tr}\left(\boldsymbol{Z}^{-1}\right) \quad \text{subject to} \tag{3.30a}$$

$$\boldsymbol{Z}_q \geq \boldsymbol{Z} \tag{3.30b}$$

$$\begin{pmatrix} \boldsymbol{G}_q^T + \boldsymbol{G}_q - \boldsymbol{Z}_q & * & * & * \\ \boldsymbol{Q}_q^{1/2} \begin{pmatrix} \boldsymbol{G}_q \\ \boldsymbol{W}_q \end{pmatrix} & \boldsymbol{I} & * & * \\ \sqrt{\gamma}\,(\boldsymbol{A}_q \boldsymbol{G}_q + \boldsymbol{B}_q \boldsymbol{W}_q) & 0 & \boldsymbol{Z}_q & * \\ \sqrt{1-\gamma}\,(\boldsymbol{A}_q \boldsymbol{G}_q + \boldsymbol{B}_q \boldsymbol{W}_q) & 0 & 0 & \boldsymbol{Z}_{q^+} \end{pmatrix} > 0 \tag{3.30c}$$

for all $q \in \mathbb{J}$ and an arbitrary $q^+ \neq q$ for each q, with the LMI variables $\boldsymbol{W}_q \in \mathbb{R}^{m \times (n+m)}$ and $\boldsymbol{G}_q \in \mathbb{R}^{(n+m) \times (n+m)}$ unrestricted and $\boldsymbol{Z}, \boldsymbol{Z}_q \in \mathbb{R}^{(n+m) \times (n+m)}$ symmetric and positive definite. The scalar variable $\gamma \in [0, 1)$ is predefined by gridding. The control gain and scheduling matrices result from

$$\boldsymbol{K}_q = \boldsymbol{W}_q \boldsymbol{G}_q^{-1} \tag{3.31a}$$

$$\boldsymbol{P}_q = \boldsymbol{Z}_q^{-1}. \tag{3.31b}$$

PROOF. The proof is divided into two parts. The first part focuses on the stability constraint (3.30c) while the second part focuses on the performance constraint (3.30b).

Assume that (3.30c) is satisfied. Then $\boldsymbol{G}_q^T + \boldsymbol{G}_q - \boldsymbol{Z}_q > \boldsymbol{0}$ and equivalently $\boldsymbol{G}_q^T + \boldsymbol{G}_q > \boldsymbol{Z}_q > \boldsymbol{0}$ holds. Thus, $\boldsymbol{G}_q$ is of full rank and therefore invertible. Since $\boldsymbol{Z}_q > \boldsymbol{0}$, also

$$(\boldsymbol{Z}_q - \boldsymbol{G}_q)^T \boldsymbol{Z}_q^{-1} (\boldsymbol{Z}_q - \boldsymbol{G}_q) \geq \boldsymbol{0} \tag{3.32}$$

holds which is equivalent to

$$\boldsymbol{G}_q^T \boldsymbol{Z}_q^{-1} \boldsymbol{G}_q \geq \boldsymbol{G}_q^T + \boldsymbol{G}_q - \boldsymbol{Z}_q. \tag{3.33}$$

Thus, (3.30c) implies

$$\begin{pmatrix} \boldsymbol{G}_q^T \boldsymbol{Z}_q^{-1} \boldsymbol{G}_q & * & * & * \\ \boldsymbol{Q}_q^{1/2} \begin{pmatrix} \boldsymbol{G}_q \\ \boldsymbol{W}_q \end{pmatrix} & \boldsymbol{I} & * & * \\ \sqrt{\gamma}\,(\boldsymbol{A}_q \boldsymbol{G}_q + \boldsymbol{B}_q \boldsymbol{W}_q) & 0 & \boldsymbol{Z}_q & * \\ \sqrt{1-\gamma}\,(\boldsymbol{A}_q \boldsymbol{G}_q + \boldsymbol{B}_q \boldsymbol{W}_q) & 0 & 0 & \boldsymbol{Z}_{q^+} \end{pmatrix} > 0. \tag{3.34}$$

Substituting $\boldsymbol{W}_q = \boldsymbol{K}_q \boldsymbol{G}_q$ and pre-/post-multiplying (3.34) by $\operatorname{diag}(\boldsymbol{G}_q^{-T}, \boldsymbol{I}, \boldsymbol{I}, \boldsymbol{I})$ and $\operatorname{diag}(\boldsymbol{G}_q^{-1}, \boldsymbol{I}, \boldsymbol{I}, \boldsymbol{I})$, respectively, results in

$$\begin{pmatrix} \boldsymbol{Z}_q^{-1} & * & * & * \\ \boldsymbol{Q}_q^{1/2} \begin{pmatrix} \boldsymbol{I} \\ \boldsymbol{K}_q \end{pmatrix} & \boldsymbol{I} & * & * \\ \sqrt{\gamma}\,(\boldsymbol{A}_q + \boldsymbol{B}_q \boldsymbol{K}_q) & 0 & \boldsymbol{Z}_q & * \\ \sqrt{1-\gamma}\,(\boldsymbol{A}_q + \boldsymbol{B}_q \boldsymbol{K}_q) & 0 & 0 & \boldsymbol{Z}_{q^+} \end{pmatrix} > 0. \tag{3.35}$$

Substituting $\boldsymbol{Z}_q = \boldsymbol{P}_q^{-1}$ and applying the Schur complement leads to

$$\boldsymbol{P}_q - \tilde{\boldsymbol{A}}_q^T(\gamma\boldsymbol{P}_q + (1-\gamma)\boldsymbol{P}_{q^+})\tilde{\boldsymbol{A}}_q - \tilde{\boldsymbol{Q}}_q > 0 \tag{3.36}$$

or equivalently

$$\tilde{\boldsymbol{A}}_q^T(\gamma\boldsymbol{P}_q + (1-\gamma)\boldsymbol{P}_{q^+})\tilde{\boldsymbol{A}}_q - \boldsymbol{P}_q < -\tilde{\boldsymbol{Q}}_q. \tag{3.37}$$

If (3.37) is feasible, also condition (3.23) of Theorem 3.1 has a solution setting $\mu_{qq} = \gamma$, $\mu_{q^+q} = 1-\gamma$, and the remaining scalars μ_{lq} zero. Consequently, the closed-loop system (3.18) is GAS.

Substituting $\boldsymbol{Z}_q = \boldsymbol{P}_q^{-1}$ as well as, pre-multiplying (3.30b) with $\boldsymbol{P}_q$ and post-multiplying (3.30b) with $\boldsymbol{Z}^{-1}$ results in

$$\boldsymbol{P}_q \leq \boldsymbol{Z}^{-1}. \tag{3.38}$$

Taking the trace of (3.38) leads to

$$\operatorname{tr}(\boldsymbol{P}_q) \leq \operatorname{tr}\left(\boldsymbol{Z}^{-1}\right). \tag{3.39}$$

Consequently, the objective function (3.30a) minimizes an upper bound on the cost function given by

$$J < \operatorname{tr}(\boldsymbol{P}_{j(0)})\,\|\boldsymbol{x}(0)\|_2^2 \leq \operatorname{tr}\left(\boldsymbol{Z}^{-1}\right)\|\boldsymbol{x}(0)\|_2^2. \tag{3.40}$$

This completes the proof. □

Remark 3.2. The usage of the more conservative stability condition (3.37) for the codesign is motivated by the fact that Theorem 3.1 is a bilinear matrix inequality problem, which is not convex. By gridding the scalar $\gamma \in [0,1)$, Theorem 3.2 becomes an LMI optimization problem for each grid point and can be solved efficiently by LMI solvers.

3.4 Observer Design

Under the assumption that only the output vector $\boldsymbol{y}(k)$ is measurable, an observer for estimating the states is necessary, such that the control law and the scheduling law can be applied based on the estimated state vector $\bar{\boldsymbol{x}}(k)$. Thus, the control law (3.16) and the scheduling law (3.17) is rewritten as

$$\boldsymbol{u}(k) = \boldsymbol{K}_{j(k)}\bar{\boldsymbol{x}}(k) \tag{3.41}$$

and

$$j(k) = \arg\min_{l\in\mathbb{J}} \bar{\boldsymbol{x}}^T(k)\boldsymbol{P}_l\bar{\boldsymbol{x}}(k). \tag{3.42}$$

However, only the physical states $\boldsymbol{x}_{\mathrm{p}i}(k)$ need to be estimated as the preceding control vector $\boldsymbol{u}_i(k-1)$ can be stored in the processor for the estimation purpose. Therefore, the estimated state vector in (3.41) and (3.42) is defined as

$$\bar{\boldsymbol{x}}(k) = \left(\begin{pmatrix}\bar{\boldsymbol{x}}_{\mathrm{p}1}(k)\\ \boldsymbol{u}_1(k-1)\end{pmatrix}^T, \ldots, \begin{pmatrix}\bar{\boldsymbol{x}}_{\mathrm{p}M}(k)\\ \boldsymbol{u}_M(k-1)\end{pmatrix}^T\right)^T. \tag{3.43}$$

In the following two types of observers are considered, namely prediction observer and current observer [FPW98, Section 8.2].

3.4.1 Prediction Observer

The structure of the prediction observer of a plant P_i is given by

$$\begin{aligned}\bar{\boldsymbol{x}}_{\mathrm{p}i}(k+1) = {} & \boldsymbol{\Phi}_{ij(k)}\bar{\boldsymbol{x}}_{\mathrm{p}i}(k) + \boldsymbol{\Gamma}_{0ij(k)}\boldsymbol{u}_i(k) + \boldsymbol{\Gamma}_{1ij(k)}\boldsymbol{u}_i(k-1) \\ & + \boldsymbol{L}_{\mathrm{P}ij(k)}\big(\boldsymbol{y}_i(k) - \boldsymbol{C}_{\mathrm{p}i}\bar{\boldsymbol{x}}_{\mathrm{p}i}(k)\big)\end{aligned} \tag{3.44}$$

with the switched observer gain matrix $\boldsymbol{L}_{\mathrm{P}ij(k)} \in \mathbb{R}^{n_i \times p_i}$, and the other matrices defined according to (3.9). Introducing the observation error

$$\tilde{\boldsymbol{x}}_{\mathrm{p}i}(k) = \boldsymbol{x}_{\mathrm{p}i}(k) - \bar{\boldsymbol{x}}_{\mathrm{p}i}(k) \tag{3.45}$$

the behavior of the observation error can be described by the difference equation

$$\tilde{\boldsymbol{x}}_{\mathrm{p}i}(k+1) = \left(\boldsymbol{\Phi}_{ij(k)} - \boldsymbol{L}_{\mathrm{P}ij(k)}\boldsymbol{C}_{\mathrm{p}i}\right)\tilde{\boldsymbol{x}}_{\mathrm{p}i}(k). \tag{3.46}$$

The behavior of the overall observation error can then be modeled with respect to the task index $j(k)$ by the block-diagonal discrete-time switched linear system

$$\tilde{\boldsymbol{x}}_{\mathrm{p}}(k+1) = \left(\boldsymbol{\Phi}_{j(k)} - \boldsymbol{L}_{\mathrm{P}j(k)}\boldsymbol{H}\right)\tilde{\boldsymbol{x}}_{\mathrm{p}}(k) \tag{3.47}$$

with

$$\tilde{\boldsymbol{x}}_{\mathrm{p}}(k) = \left(\tilde{\boldsymbol{x}}_{\mathrm{p}1}^T(k) \quad \cdots \quad \tilde{\boldsymbol{x}}_{\mathrm{p}M}^T(k)\right)^T \tag{3.48a}$$

$$\boldsymbol{\Phi}_{j(k)} = \mathrm{diag}\left(\boldsymbol{\Phi}_{1j(k)}, \ldots, \boldsymbol{\Phi}_{Mj(k)}\right) \tag{3.48b}$$

$$\boldsymbol{L}_{\mathrm{P}j(k)} = \mathrm{diag}\left(\boldsymbol{L}_{\mathrm{P}1j(k)}, \ldots, \boldsymbol{L}_{\mathrm{P}Mj(k)}\right) \tag{3.48c}$$

$$\boldsymbol{H} = \mathrm{diag}\left(\boldsymbol{C}_{\mathrm{p}1}, \ldots, \boldsymbol{C}_{\mathrm{p}M}\right). \tag{3.48d}$$

As a performance measure for the observer design, a decay rate $\rho \in \mathbb{R}^+$ is defined such that the closed-loop observation error dynamic (3.46) is globally exponentially stable (GES).

Definition 3.1 *[Kha02, Section 4.5] The switched linear system* (3.47) *is said to be globally exponentially stable (GES) with a decay rate ρ if there exists a constant $c \in \mathbb{R}^+$ such that*

$$\|\tilde{\boldsymbol{x}}_{\mathrm{p}}(k)\|_2 \leq c e^{-\rho \sum_{l=0}^{k-1} h_{j(l)}} \|\tilde{\boldsymbol{x}}_{\mathrm{p}}(0)\|_2 \tag{3.49}$$

for all $k \in \mathbb{N}$ for any initial state $\tilde{\boldsymbol{x}}_{\mathrm{p}}(0)$.

Problem 3.2 for the observer design can be now reformulated as

Problem 3.4 *Find the observer gain matrix $\boldsymbol{L}_{\mathrm{P}i}$ for all $i \in \mathbb{J}$ such that closed-loop switched linear system* (3.47) *is GES with the decay rate ρ.*

The observer design is conducted assuming the switched system (3.47) is an arbitrary switched system, where stability can be analyzed based on a common quadratic Lyapunov function [Kan97, LA09]. Stability is then guaranteed under any arbitrary switching sequence and thus, also under the switching law (3.42). The common quadratic Lyapunov function is defined as

$$V_{\mathrm{e}}(k) = \tilde{\boldsymbol{x}}_{\mathrm{p}}^{T}(k)\boldsymbol{S}\tilde{\boldsymbol{x}}_{\mathrm{p}}(k) \tag{3.50}$$

with $\boldsymbol{S} \in \mathbb{R}^{n\times n}$ symmetric and positive definite.

Theorem 3.3 *The solution to Problem 3.4 is obtained from the LMI feasibility problem*

$$\begin{pmatrix} \boldsymbol{S}e^{-2\rho h_q} & * \\ \boldsymbol{S}\boldsymbol{\Phi}_q - \boldsymbol{F}_q\boldsymbol{H} & \boldsymbol{S} \end{pmatrix} > \boldsymbol{0} \tag{3.51}$$

for all scheduling indices $q \in \mathbb{J}$ and a desired decay rate $\rho > 0$ with the LMI variables $\boldsymbol{F}_q \in \mathbb{R}^{n\times p}$ unrestricted and $\boldsymbol{S} \in \mathbb{R}^{n\times n}$ symmetric and positive definite. The observer gain matrices result from

$$\boldsymbol{L}_{\mathrm{P}q} = \boldsymbol{S}^{-1}\boldsymbol{F}_q. \tag{3.52}$$

Proof. Assume that (3.51) is satisfied. Substituting $\boldsymbol{F}_q = \boldsymbol{S}\boldsymbol{L}_{\mathrm{P}q}$ and pre-/post-multiplying (3.51) by $\mathrm{diag}(\boldsymbol{I}, \boldsymbol{S}^{-1})$ and its transpose, respectively, results in

$$\begin{pmatrix} \boldsymbol{S}e^{-2\rho h_q} & * \\ \boldsymbol{\Phi}_q - \boldsymbol{L}_{\mathrm{P}q}\boldsymbol{H} & \boldsymbol{S}^{-1} \end{pmatrix} > 0. \tag{3.53}$$

Applying the Schur complement leads to

$$\boldsymbol{S}e^{-2\rho h_q} - \left(\boldsymbol{\Phi}_q - \boldsymbol{L}_{\mathrm{P}q}\boldsymbol{H}\right)^T \boldsymbol{S}\left(\boldsymbol{\Phi}_q - \boldsymbol{L}_{\mathrm{P}q}\boldsymbol{H}\right) > 0. \tag{3.54}$$

Assuming that q denotes the scheduling index at a time instant t_k, i.e. $j(k) = q$, pre-/post-multiplying (3.54) with $\tilde{\boldsymbol{x}}_{\mathrm{p}}^{T}(k)$ and its transpose, respectively, results in

$$V_{\mathrm{e}}(k+1) < V_{\mathrm{e}}(k)e^{-2\rho h_{j(k)}}. \tag{3.55}$$

Based on (3.55) it can be concluded that

$$V_{\mathrm{e}}(k) < V_{\mathrm{e}}(0)e^{-2\rho\sum_{l=0}^{k-1} h_{j(l)}} \tag{3.56}$$

holds. Based on the definition of the Lyapunov function (3.50), there exist positive constants $\beta_1 = \lambda_{\min}(\boldsymbol{S})$ and $\beta_2 = \lambda_{\max}(\boldsymbol{S})$ such that

$$\beta_1 \left\|\tilde{\boldsymbol{x}}_{\mathrm{p}}(k)\right\|_2^2 \leq V_{\mathrm{e}}(k) \leq \beta_2 \left\|\tilde{\boldsymbol{x}}_{\mathrm{p}}(k)\right\|_2^2 \tag{3.57}$$

holds for any $\tilde{\boldsymbol{x}}_{\mathrm{p}}(k) \in \mathbb{R}^n$. Hence,

$$\begin{aligned} \left\|\tilde{\boldsymbol{x}}_{\mathrm{p}}(k)\right\|_2 &\leq \left(\frac{V_{\mathrm{e}}(k)}{\beta_1}\right)^{1/2} < \left(\frac{V_{\mathrm{e}}(0)e^{-2\rho\sum_{l=0}^{k-1} h_{j(l)}}}{\beta_1}\right)^{1/2} \\ &\leq \left(\frac{\beta_2 \left\|\tilde{\boldsymbol{x}}_{\mathrm{p}}(0)\right\|_2^2 e^{-2\rho\sum_{l=0}^{k-1} h_{j(l)}}}{\beta_1}\right)^{1/2} = \left(\frac{\beta_2}{\beta_1}\right)^{1/2} \left\|\tilde{\boldsymbol{x}}_{\mathrm{p}}(0)\right\|_2 e^{-\rho\sum_{l=0}^{k-1} h_{j(l)}} \end{aligned}$$

such that the closed-loop system (3.47) is GES with the decay rate ρ. This completes the proof. □

3.4.2 Current Observer

One drawback of the prediction observer is that the estimated state vector $\bar{\boldsymbol{x}}(k)$ is determined based on the precedent measurement $\boldsymbol{y}(k-1)$. To overcome this drawback, a current observer is designed, where the state vector is estimated based on the current measurement $\boldsymbol{y}(k)$. The structure of the current observer of a plant P_i is given by

$$\bar{\boldsymbol{x}}_{\mathrm{p}i}(k) = \check{\boldsymbol{x}}_{\mathrm{p}i}(k) + \boldsymbol{L}_{\mathrm{C}ij(k-1)}\big(\boldsymbol{y}_i(k) - \boldsymbol{C}_{\mathrm{p}i}\check{\boldsymbol{x}}_{\mathrm{p}i}(k)\big) \tag{3.58}$$

where

$$\check{\boldsymbol{x}}_{\mathrm{p}i}(k) = \boldsymbol{\Phi}_{ij(k-1)}\bar{\boldsymbol{x}}_{\mathrm{p}i}(k-1) + \boldsymbol{\Gamma}_{0ij(k-1)}\boldsymbol{u}_i(k-1) + \boldsymbol{\Gamma}_{1ij(k-1)}\boldsymbol{u}_i(k-2). \tag{3.59}$$

For describing the behavior of the observation error (3.45), equation (3.59) is substituted into (3.58) leading to the error dynamic

$$\tilde{\boldsymbol{x}}_{\mathrm{p}i}(k+1) = \left(\boldsymbol{\Phi}_{ij(k)} - \boldsymbol{L}_{\mathrm{C}ij(k)}\boldsymbol{C}_{\mathrm{p}i}\boldsymbol{\Phi}_{ij(k)}\right)\tilde{\boldsymbol{x}}_{\mathrm{p}i}(k). \tag{3.60}$$

The overall observation error dynamic is then modeled with respect to the task index $j(k)$ by the block-diagonal discrete-time switched linear system

$$\tilde{\boldsymbol{x}}_{\mathrm{p}}(k+1) = \left(\boldsymbol{\Phi}_{j(k)} - \boldsymbol{L}_{\mathrm{C}j(k)}\boldsymbol{H}\boldsymbol{\Phi}_{j(k)}\right)\tilde{\boldsymbol{x}}_{\mathrm{p}}(k) \tag{3.61}$$

with the variables defined according to (3.48). Problem 3.2 for the current observer design is reformulated as

Problem 3.5 *Find the observer gain matrix $\boldsymbol{L}_{\mathrm{C}i}$ for all $i \in \mathbb{J}$ such that closed-loop switched linear system (3.61) is GES with the decay rate ρ.*

Analog to the prediction observer, a theorem for designing the observer gain matrix is proposed based on the common quadratic Lyapunov function (3.50).

Theorem 3.4 *The solution to Problem 3.5 is obtained from the LMI feasibility problem*

$$\begin{pmatrix} \boldsymbol{S}e^{-2\rho h_q} & * \\ \boldsymbol{S}\boldsymbol{\Phi}_q - \boldsymbol{F}_q\boldsymbol{H}\boldsymbol{\Phi}_q & \boldsymbol{S} \end{pmatrix} > 0 \tag{3.62}$$

for all scheduling indices $q \in \mathbb{J}$ and a desired decay rate $\rho > 0$ with the LMI variables $\boldsymbol{F}_q \in \mathbb{R}^{n\times p}$ unrestricted and $\boldsymbol{S} \in \mathbb{R}^{n\times n}$ symmetric and positive definite. The observer gain matrices result from

$$\boldsymbol{L}_{\mathrm{C}q} = \boldsymbol{S}^{-1}\boldsymbol{F}_q. \tag{3.63}$$

PROOF. The proof of Theorem 3.4 follows the lines of the proof of Theorem 3.3 by substituting $\boldsymbol{H} := \boldsymbol{H}\boldsymbol{\Phi}_q$. □

3.5 Separation Principle for Switched Systems

In the proposed framework the output-feedback control and scheduling problem is decomposed into two subproblems (Problem 3.1 and 3.2). The control and scheduling codesign is derived under the assumption that the true state vector $\boldsymbol{x}(k)$ is fed back. Therefore, it is necessary to investigate the effect of the state estimation on the system dynamics considering stability and performance. It is analyzed if the classical separation principle still holds. For simplicity the derivations are given for the prediction observer. For the current observer, it follows the same lines.

3.5.1 Stability Analysis

In the following, the stability of the complete system including the controlled plants and the observation error is analyzed.

Theorem 3.5 *Applying the prediction observer* (3.44)*, the scheduling law* (3.42)*, and the control law* (3.41) *with the observer gain matrices determined based on Theorem 3.3 and the control and scheduling matrices determined based on Theorem 3.2, the discrete-time switched linear system* (3.10) *is globally asymptotically stabilized.*

PROOF. First, the closed-loop system composed of (3.10), the control law (3.41), the scheduling law (3.42) and the observer (3.44) is written as

$$\begin{pmatrix} \bar{\boldsymbol{x}}(k+1) \\ \tilde{\boldsymbol{x}}_{\mathrm{p}}(k+1) \end{pmatrix} = \begin{pmatrix} \boldsymbol{A}_{j(k)} + \boldsymbol{B}_{j(k)}\boldsymbol{K}_{j(k)} & \boldsymbol{M}\boldsymbol{L}_{\mathrm{P}j(k)}\boldsymbol{H} \\ 0 & \boldsymbol{\Phi}_{j(k)} - \boldsymbol{L}_{\mathrm{P}j(k)}\boldsymbol{H} \end{pmatrix} \begin{pmatrix} \bar{\boldsymbol{x}}(k) \\ \tilde{\boldsymbol{x}}_{\mathrm{p}}(k) \end{pmatrix} \tag{3.64}$$

with $\boldsymbol{M} = \mathrm{diag}\left(\binom{I}{0}, \ldots, \binom{I}{0}\right) \in \mathbb{R}^{(n+m)\times n}$ and the other matrices defined according to (3.12), (3.43) and (3.48). Further, consider the Lyapunov function

$$V(k) = \min_{l\in\mathbb{J}} \begin{pmatrix} \bar{\boldsymbol{x}}(k) \\ \tilde{\boldsymbol{x}}_{\mathrm{p}}(k) \end{pmatrix}^T \begin{pmatrix} \boldsymbol{P}_l & 0 \\ 0 & \theta\boldsymbol{S} \end{pmatrix} \begin{pmatrix} \bar{\boldsymbol{x}}(k) \\ \tilde{\boldsymbol{x}}_{\mathrm{p}}(k) \end{pmatrix} \tag{3.65}$$

where the scalar $\theta \in \mathbb{R}^+$ is to be determined and the matrices $\boldsymbol{P}_l$ and $\boldsymbol{S}$ result from Theorem 3.2 and Theorem 3.3, respectively. The Lyapunov function (3.65) can equivalently be written as

$$V(k) = \theta\tilde{\boldsymbol{x}}_{\mathrm{p}}^T(k)\boldsymbol{S}\tilde{\boldsymbol{x}}_{\mathrm{p}}(k) + \min_{l\in\mathbb{J}} \bar{\boldsymbol{x}}^T(k)\boldsymbol{P}_l\bar{\boldsymbol{x}}(k). \tag{3.66}$$

As the first term on the right-hand side of (3.66) does not affect the minimization operation, the argument of the Lyapunov function (3.65) is equivalent to the scheduling index resulting from the scheduling law (3.42). Assume that the scheduling index at a time instant t_k is $j(k) = q$. Then, the difference of the Lyapunov function

$\Delta V(k) = V(k+1) - V(k)$ along the closed-loop trajectories is given by

$$\Delta V(k) = \min_{l\in\mathbb{J}} \begin{pmatrix} \bar{\boldsymbol{x}}(k) \\ \tilde{\boldsymbol{x}}_{\mathrm{p}}(k) \end{pmatrix}^T \begin{pmatrix} \tilde{\boldsymbol{A}}_q^T \boldsymbol{P}_l \tilde{\boldsymbol{A}}_q - \boldsymbol{P}_q & \tilde{\boldsymbol{A}}_q^T \boldsymbol{P}_l \boldsymbol{M}\boldsymbol{L}_{\mathrm{P}_q}\boldsymbol{H} \\ * & \theta(\tilde{\boldsymbol{\Phi}}_q^T \boldsymbol{S}\tilde{\boldsymbol{\Phi}}_q - \boldsymbol{S}) + (\boldsymbol{M}\boldsymbol{L}_{\mathrm{P}_q}\boldsymbol{H})^T \boldsymbol{P}_l (\boldsymbol{M}\boldsymbol{L}_{\mathrm{P}_q}\boldsymbol{H}) \end{pmatrix} \begin{pmatrix} \bar{\boldsymbol{x}}(k) \\ \tilde{\boldsymbol{x}}_{\mathrm{p}}(k) \end{pmatrix}$$

$$\leq \begin{pmatrix} \bar{\boldsymbol{x}}(k) \\ \tilde{\boldsymbol{x}}_{\mathrm{p}}(k) \end{pmatrix}^T \bar{\boldsymbol{\mathcal{P}}}_q \begin{pmatrix} \bar{\boldsymbol{x}}(k) \\ \tilde{\boldsymbol{x}}_{\mathrm{p}}(k) \end{pmatrix} \tag{3.67}$$

with

$$\bar{\boldsymbol{\mathcal{P}}}_q = \begin{pmatrix} \tilde{\boldsymbol{A}}_q^T \boldsymbol{P}_{qq^+} \tilde{\boldsymbol{A}}_q - \boldsymbol{P}_q & \tilde{\boldsymbol{A}}_q^T \boldsymbol{P}_{qq^+} \boldsymbol{M}\boldsymbol{L}_{\mathrm{P}_q}\boldsymbol{H} \\ * & (\boldsymbol{M}\boldsymbol{L}_{\mathrm{P}_q}\boldsymbol{H})^T \boldsymbol{P}_{qq^+} (\boldsymbol{M}\boldsymbol{L}_{\mathrm{P}_q}\boldsymbol{H}) + \theta(\tilde{\boldsymbol{\Phi}}_q^T \boldsymbol{S}\tilde{\boldsymbol{\Phi}}_q - \boldsymbol{S}) \end{pmatrix} \tag{3.68}$$

where $\tilde{\boldsymbol{\Phi}}_q = \boldsymbol{\Phi}_q - \boldsymbol{L}_{\mathrm{P}_q}\boldsymbol{H}$ and $\boldsymbol{P}_{qq^+} = \gamma\boldsymbol{P}_q + (1-\gamma)\boldsymbol{P}_{q^+}$ result from Theorem 3.2. Based on Theorem 3.2 and Theorem 3.3 it is known that the inequalities

$$\tilde{\boldsymbol{A}}_q^T \boldsymbol{P}_{qq^+} \tilde{\boldsymbol{A}}_q - \boldsymbol{P}_q < 0 \tag{3.69}$$

$$\tilde{\boldsymbol{\Phi}}_q^T \boldsymbol{S}\tilde{\boldsymbol{\Phi}}_q - \boldsymbol{S} < 0 \tag{3.70}$$

hold for all $q \in \mathbb{J}$. By selecting $\theta > 0$ large enough, $\bar{\boldsymbol{\mathcal{P}}}_q$ is negative definite for all $q \in \mathbb{J}$, such that $\Delta V(k)$ is negative definite, see also the discussions in [Zha01, BMM09]. Thus, the closed-loop system (3.64) is GAS.

The state vector $\boldsymbol{x}(k)$ can be extracted from the estimated state vector $\bar{\boldsymbol{x}}(k)$ and the observation error $\tilde{\boldsymbol{x}}_{\mathrm{p}}(k)$ using the relationship

$$\boldsymbol{x}(k) = \bar{\boldsymbol{x}}(k) + \boldsymbol{M}\tilde{\boldsymbol{x}}_{\mathrm{p}}(k). \tag{3.71}$$

Hence, the convergence of $\bar{\boldsymbol{x}}(k)$ and $\tilde{\boldsymbol{x}}_{\mathrm{p}}(k)$ towards the origin implies the convergence of $\boldsymbol{x}(k)$. This completes the proof. □

3.5.2 Performance Analysis

In this section it is analyzed how the state estimation affects the guaranteed performance introduced for the control and scheduling design. Considering the complete closed-loop system model (3.64), also the cost function (3.15) is rewritten with respect to the estimated state vector $\bar{\boldsymbol{x}}(k)$ and the observation error $\tilde{\boldsymbol{x}}_{\mathrm{p}}(k)$. Substituting (3.41) and (3.71) into (3.15) results in

$$J = \sum_{k=0}^{\infty} \begin{pmatrix} \bar{\boldsymbol{x}}(k) \\ \tilde{\boldsymbol{x}}_{\mathrm{p}}(k) \end{pmatrix}^T \bar{\boldsymbol{\mathcal{Q}}}_{j(k)} \begin{pmatrix} \bar{\boldsymbol{x}}(k) \\ \tilde{\boldsymbol{x}}_{\mathrm{p}}(k) \end{pmatrix} \tag{3.72}$$

with the positive semidefinite matrix

$$\bar{\boldsymbol{\mathcal{Q}}}_{j(k)} = \begin{pmatrix} \boldsymbol{I} & \boldsymbol{M} \\ \boldsymbol{K}_{j(k)} & \boldsymbol{0} \end{pmatrix}^T \boldsymbol{Q}_{j(k)} \begin{pmatrix} \boldsymbol{I} & \boldsymbol{M} \\ \boldsymbol{K}_{j(k)} & \boldsymbol{0} \end{pmatrix}$$

$$= \begin{pmatrix} \left(\begin{smallmatrix} \boldsymbol{I} \\ \boldsymbol{K}_{j(k)} \end{smallmatrix}\right)^T \boldsymbol{Q}_{j(k)} \left(\begin{smallmatrix} \boldsymbol{I} \\ \boldsymbol{K}_{j(k)} \end{smallmatrix}\right) & \left(\begin{smallmatrix} \boldsymbol{I} \\ \boldsymbol{K}_{j(k)} \end{smallmatrix}\right)^T \boldsymbol{Q}_{j(k)} \left(\begin{smallmatrix} \boldsymbol{M} \\ \boldsymbol{0} \end{smallmatrix}\right) \\ \left(\begin{smallmatrix} \boldsymbol{M} \\ \boldsymbol{0} \end{smallmatrix}\right)^T \boldsymbol{Q}_{j(k)} \left(\begin{smallmatrix} \boldsymbol{I} \\ \boldsymbol{K}_{j(k)} \end{smallmatrix}\right) & \left(\begin{smallmatrix} \boldsymbol{M} \\ \boldsymbol{0} \end{smallmatrix}\right)^T \boldsymbol{Q}_{j(k)} \left(\begin{smallmatrix} \boldsymbol{M} \\ \boldsymbol{0} \end{smallmatrix}\right) \end{pmatrix}. \tag{3.73}$$

Assuming the scheduling index at a time instant t_k is $j(k) = q$ the matrix

$$\bar{\mathcal{P}}_q + \bar{\mathcal{Q}}_q = \begin{pmatrix} \tilde{A}_q^T P_{qq^+} \tilde{A}_q - P_q + \tilde{Q}_q & \tilde{A}_q^T P_{qq^+} M L_{\mathrm{P}_q} H + \left(\begin{smallmatrix} I \\ K_q \end{smallmatrix}\right)^T Q_q \left(\begin{smallmatrix} M \\ 0 \end{smallmatrix}\right) \\ * & \bar{\mathcal{P}}_{22q} + \bar{\mathcal{Q}}_{22q} \end{pmatrix} \tag{3.74}$$

with

$$\bar{\mathcal{P}}_{22q} + \bar{\mathcal{Q}}_{22q} = \theta(\tilde{\Phi}_q^T S \tilde{\Phi}_q - S) + (M L_{\mathrm{P}_q} H)^T P_{qq^+} (M L_{\mathrm{P}_q} H) + \left(\begin{smallmatrix} M \\ 0 \end{smallmatrix}\right)^T Q_q \left(\begin{smallmatrix} M \\ 0 \end{smallmatrix}\right) \tag{3.75}$$

is analyzed. First, the control and scheduling parameters have been designed such that

$$\tilde{A}_q^T P_{qq^+} \tilde{A}_q - P_q + \tilde{Q}_q < 0 \tag{3.76}$$

is satisfied for all $q \in \mathbb{J}$, see inequality (3.37). Second, by chosing θ large enough both $\bar{\mathcal{P}}_{22q} + \bar{\mathcal{Q}}_{22q} < \mathbf{0}$ and $\bar{\mathcal{P}}_q + \bar{\mathcal{Q}}_q < \mathbf{0}$ hold for all $q \in \mathbb{J}$. Thus, the difference of the Lyapunov function is upper bounded by

$$\Delta V(k) \leq \begin{pmatrix} \bar{x}(k) \\ \tilde{x}_{\mathrm{p}}(k) \end{pmatrix}^T \bar{\mathcal{P}}_q \begin{pmatrix} \bar{x}(k) \\ \tilde{x}_{\mathrm{p}}(k) \end{pmatrix} < - \begin{pmatrix} \bar{x}(k) \\ \tilde{x}_{\mathrm{p}}(k) \end{pmatrix}^T \bar{\mathcal{Q}}_q \begin{pmatrix} \bar{x}(k) \\ \tilde{x}_{\mathrm{p}}(k) \end{pmatrix}. \tag{3.77}$$

Summing up (3.77) over $k = 0, ..., \infty$ results in

$$V(0) - \lim_{k \to \infty} V(k) > J. \tag{3.78}$$

As the state vector $\left(\bar{x}^T(k) \quad \tilde{x}_{\mathrm{p}}^T(k)\right)^T$ converges to the origin, $V(k) \to 0$ for $k \to \infty$, such that an upper bound can be given by

$$J < \min_{l \in \mathbb{J}} \begin{pmatrix} \bar{x}(0) \\ \tilde{x}_{\mathrm{p}}(0) \end{pmatrix}^T \begin{pmatrix} P_l & 0 \\ 0 & \theta S \end{pmatrix} \begin{pmatrix} \bar{x}(0) \\ \tilde{x}_{\mathrm{p}}(0) \end{pmatrix} < \operatorname{tr}\left(\begin{pmatrix} Z^{-1} & 0 \\ 0 & \theta S \end{pmatrix}\right) \left\| \begin{pmatrix} \bar{x}(0) \\ \tilde{x}_{\mathrm{p}}(0) \end{pmatrix} \right\|_2^2. \tag{3.79}$$

It is shown that a certain performance can be guaranteed, however the given upper bound (3.79) may be pretty conservative as the separation principle is applied for designing the controller and scheduler on the one hand and the observer on the other hand. The joint design of controller, scheduler and observer in a codesign framework remains an open problem.

3.6 Illustrative Example and Comparison

For evaluating the effectiveness of the proposed output-feedback control and scheduling design by simulation, the presented approach is applied for the simultaneous stabilization of a set of inverted pendulums, see Figure 3.2. The linearized dynamic model of each

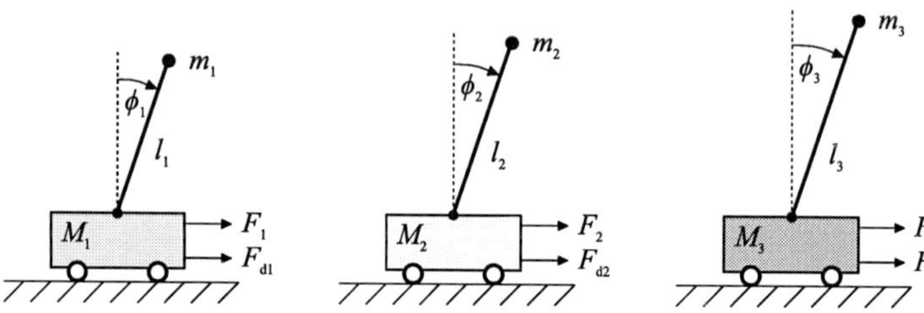

Figure 3.2: Simultaneous stabilization of three inverted pendulums

inverted pendulum is given by

$$\begin{pmatrix} \dot{\phi}_i(t) \\ \ddot{\phi}_i(t) \end{pmatrix} = \begin{pmatrix} 0 & 1 \\ \frac{(m_i+M_i)g}{M_i l_i} & 0 \end{pmatrix} \begin{pmatrix} \phi_i(t) \\ \dot{\phi}_i(t) \end{pmatrix} + \begin{pmatrix} 0 \\ \frac{-1}{M_i l_i} \end{pmatrix} (F_i(t-\tau_i) + F_{\mathrm{d}i}(t)) \tag{3.80}$$

where ϕ_i is the pendulum angle in radian, F_i is the force acting on the cart in Newton, and $F_{\mathrm{d}i}$ is a force disturbance in Newton, with $i = \{1, 2, 3\}$. The inverted pendulums have the same pendulum mass $m_i = 0.1\,\mathrm{kg}$ (concentrated at the tip) and the same cart mass $M_i = 0.1\,\mathrm{kg}$, but different pendulum lengths $l_1 = 0.136\,\mathrm{m}$, $l_2 = 0.242\,\mathrm{m}$, and $l_3 = 0.545\,\mathrm{m}$. Gravitational acceleration is considered equal to $g = 9.81\,\mathrm{m/s^2}$.

The discretization intervals and input delays are selected as

$$\begin{aligned} h_1 &= 8\,\mathrm{ms}, & h_2 &= 10\,\mathrm{ms}, & h_3 &= 12\,\mathrm{ms}, \\ \tau_1 &= 5\,\mathrm{ms}, & \tau_2 &= 5\,\mathrm{ms}, & \tau_3 &= 5\,\mathrm{ms} \end{aligned}$$

for all approaches considered in this evaluation. The scheduling overhead, which is relevant especially for online scheduling, is neglected in the simulation, assuming a processor with high performance. A more detailed discussion on the scheduling overhead is given in the experimental study in Section 3.7.

The weighting matrices of the cost function (3.2) are given by

$$\boldsymbol{Q}_{ci} = \begin{pmatrix} 1000 & 0 \\ 0 & 10 \end{pmatrix}, \quad \boldsymbol{R}_{ci} = 1$$

for all $i \in \{1, 2, 3\}$. The decay rate for the design of both observers is selected as $\rho = 150$ after several design iterations.

The force disturbance $F_{\mathrm{d}i}$ appears as a random rectangular impulse. The duration of the impulse is set $2\,\mathrm{ms}$ and the magnitude is uniformly distributed in the interval $[-50\,\mathrm{N}, 50\,\mathrm{N}]$. The duration between two impulses is uniformly distributed in the interval $[0.4\,\mathrm{s}, 1.7\,\mathrm{s}]$.

3.6.1 State-Feedback Offline and Online Scheduling

In the following a comparison of the proposed state-feedback online scheduling approach (SF-on) with a state-feedback offline scheduling (SF-off) approach is made to show the benefits of the online scheduling. For the state-feedback offline scheduling approach the network communication is scheduled under the periodic sequence $\mathcal{S} := (1, 2, 3)$, i.e. $j(k) = (k \bmod 3) + 1$ with $k \in \mathbb{N}_0$. It is assumed that all states are measurable. The control parameters are then computed based on the periodic optimal control method proposed in [GIL07] taking the input delay into account. Thereby, the periodic control problem is transformed into a time-invariant optimization problem [BCD90, BCD91].

For the online scheduling approach the control and scheduling parameters are determined using Theorem 3.2, which is implemented in MATLAB using the toolbox CVX [GB12, GB08] and the SeDuMi solver [Stu99]. For simplicity, the index q^+ in (3.30c) is set

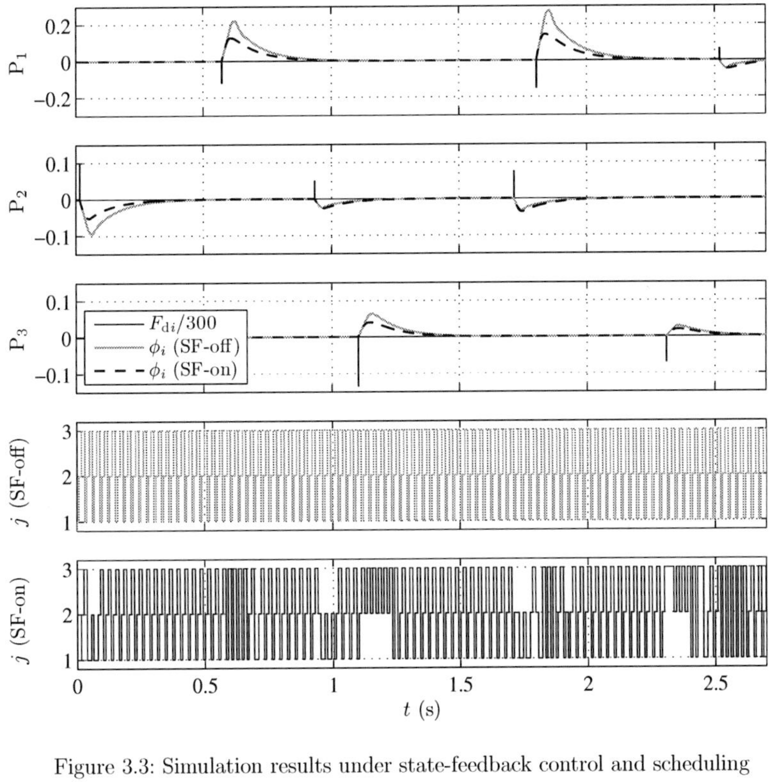

Figure 3.3: Simulation results under state-feedback control and scheduling

$q^+ = (q \bmod 3) + 1$ for each $q \in \mathbb{J}$. The minimal objective function $\operatorname{tr}(\boldsymbol{Z}^{-1})$ is found for $\gamma = 0$. Under the random force disturbance both the state-feedback offline and online scheduling approach are applied with the simulation time $T_{\text{sim}} = 100\,\text{s}$. The simulation is realized with MATLAB/Simulink using the toolbox TrueTime [OHC07, CHL+03].

Figure 3.3 shows an excerpt of the simulation results for $t \in [0.0\,\text{s}, 2.7\,\text{s}]$. The scheduling index $j(k)$ shows that under online scheduling the control signal of a plant P_i is more frequently updated if the plant is subject to disturbances. This can be clearly seen in the interval $t \in [1.1\,\text{s}, 1.25\,\text{s}]$ where only the control input u_2 and u_3 are updated as the corresponding plants are subject to disturbances. Besides a more frequent update of the control input of a disturbed plant the online scheduling allows a more immediate reaction to disturbances. This leads to an essentially better rejection of the disturbances as can be seen in the interval $t \in [0.57\,\text{s}, 0.8\,\text{s}]$ and $t \in [1.8\,\text{s}, 2.1\,\text{s}]$ for plant P_1 or in the interval $t \in [0.01\,\text{s}, 0.3\,\text{s}]$ for plant P_2. Therefore, the cost under online scheduling $J_{\text{SF-on}} = 559.3$ is smaller than under offline scheduling $J_{\text{SF-off}} = 973.5$.

Remark 3.3. In the time interval $t \in [0\,\text{s}, 0.5\,\text{s}]$, only the plant P_2 is disturbed. Therefore, another feasible online scheduling may consist in controlling only plant P_2 from the time instant of the disturbance on plant P_2 ($t = 0.015\,\text{s}$) until another plant is disturbed.

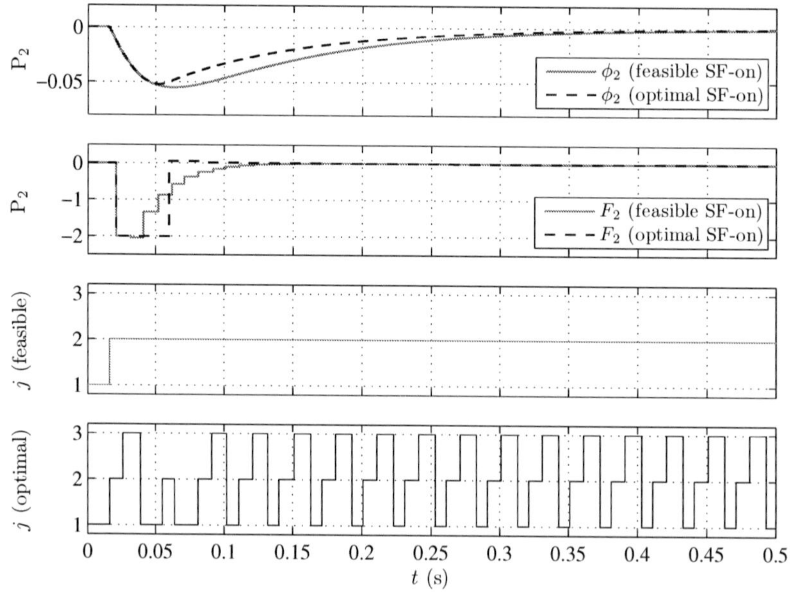

Figure 3.4: Simulation results under different online scheduling approaches

The feasible online scheduling is shown in the third plot of Figure 3.4. The control parameters are equivalent to the ones received from Theorem 3.2. The output angle ϕ_2, which is given in Figure 3.4, shows that this leads to a worse performance than the proposed online scheduling law (3.17). This can be explained by the fact that the online scheduling law (3.17) selects the scheduling index $j(k)$, such that the optimal performance is achieved under the given possibilities. The simulation demonstrates that holding the previous control signal may lead to a better control performance than updating the control signal, which is considered by the scheduling law (3.17), see also the second plot in Figure 3.4, where the control input F_2 is shown. Therefore, the scheduling law (3.17) may decide to update the control input of a plant which is in the steady state instead of the disturbed plant to realize the optimal control performance. This is also reflected by the cost J. After the simulation time $T_{\text{sim}} = 0.5\,\text{s}$, the cost is $J_{\text{optimal SF-on}} = 1.4326$ under the optimal scheduling law (3.17) and $J_{\text{feasible SF-on}} = 1.5006$ under the feasible online scheduling shown in the third plot of Figure 3.4.

3.6.2 Output-Feedback Online Scheduling

Analog to Section 3.6.1 the output-feedback online scheduling approaches are evaluated with the pendulum angle ϕ_i as output, i.e. $\boldsymbol{C}_{\text{p}i} = \begin{pmatrix} 1 & 0 \end{pmatrix}$ for all $i \in \mathbb{J}$. Additional to the control and scheduling parameters the observer gain matrices are designed using Theorem 3.3 and 3.4, respectively, which are implemented using the MATLAB toolbox YALMIP [Löf04] and solved using the SeDuMi solver [Stu99].

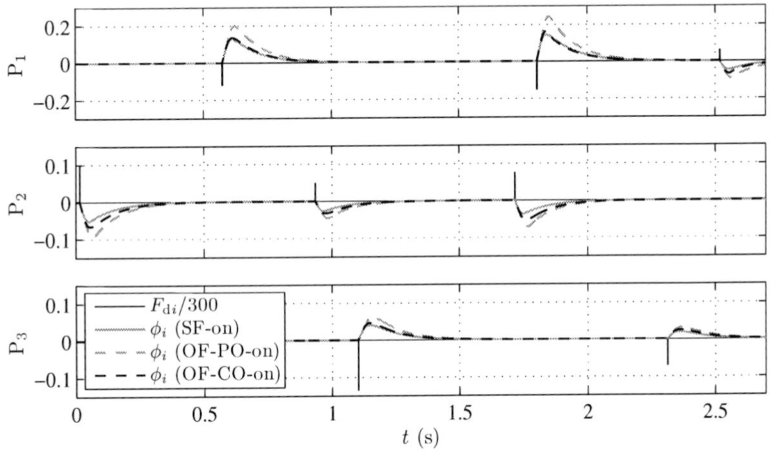

Figure 3.5: Simulation results under output-feedback control and scheduling

An excerpt of the simulation results is shown in Figure 3.5 for $t \in [0.0\,\mathrm{s}, 2.7\,\mathrm{s}]$. For all plants the output behavior is given under state-feedback online scheduling (SF-on), output-feedback online scheduling with prediction observer (OF-PO-on) and output-feedback online scheduling with current observer (OF-CO-on). Under the output-feedback strategy with prediction observer the disturbances are rejected worst for all plants leading to the highest cost, see Table 3.1. For plant P_1 the disturbance rejection under the output-feedback strategy with current observer works similar well as under state-feedback online scheduling in the shown excerpt in Figure 3.5. However, for the other plants a small performance degradation compared with the state-feedback online scheduling strategy can be seen resulting to higher costs, as indicated in Table 3.1.

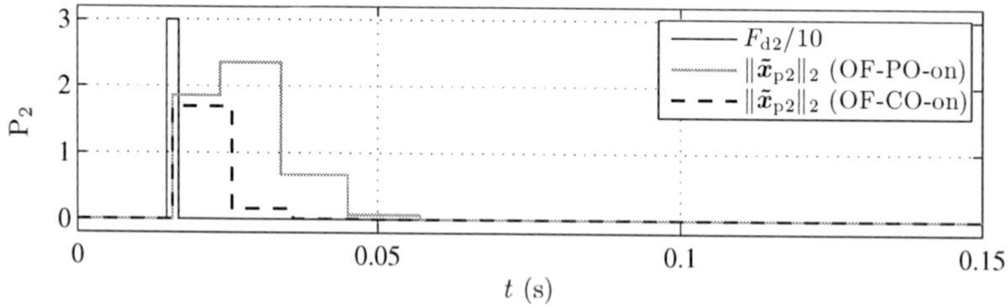

Figure 3.6: Norm of observation error $\|\tilde{\boldsymbol{x}}_{\mathrm{p}2}\|_2$ of plant P_2

The reason for such performance degradation under the output-feedback strategies lies essentially in the observation error. As both the control input and the scheduling index are computed based on the estimated states, which include an observation error and therefore maybe defective, a certain performance degradation is unavoidable. Figure 3.6 shows by means of plant P_2 that the current observer can correct the observation error faster than the prediction observer resulting in a smaller norm of the observation error.

Method	J
SF-off	973.5
SF-on	559.3
OF-PO-on	1173.8
OF-CO-on	736.4

Table 3.1: Costs under different scheduling strategies

Table 3.1 summarizes the results of the discussed state-feedback and output-feedback control and scheduling approaches. The state-feedback online scheduling strategy clearly shows the best performance, whereas the output-feedback approach with prediction observer has the worst performance. It is interesting that under output-feedback control and scheduling with current observer a better performance than under state-feedback offline scheduling can be realized. Thus, the proposed output-feedback control and scheduling approach with current observer is a promising strategy if not all states are measurable.

3.7 Experimental Studies

Besides the simulative evaluation the proposed online scheduling strategy is evaluated experimentally based on two setups, namely the simultaneous control of two DC motors and one inverted pendulum (Figure 3.7) and the simultaneous control of two double integrators (Figure 3.10). In this experimental study the implementability, especially of the scheduler, on a microcontroller and the performance in a practical environment is analyzed. As network imperfection, such as time-varying delay, packet loss and quantization effects are not considered in the proposed method, the control signals are sent hardwired to the actuator for simplicity.

3.7.1 Two DC Motors and one Inverted Pendulum

In the first experiment the state-feedback offline and online scheduling are compared for the simultaneous control of two DC motors (Maxon RE-max29) with an attached wheel and one inverted pendulum on a cart by one microcontroller NXP LPC2294. For the DC motors (plant P_1 and P_2) the control aim is to track a constant reference angle r_i and for the inverted pendulum P_3 both the angle ϕ_3 and the position p shall be stabilized.

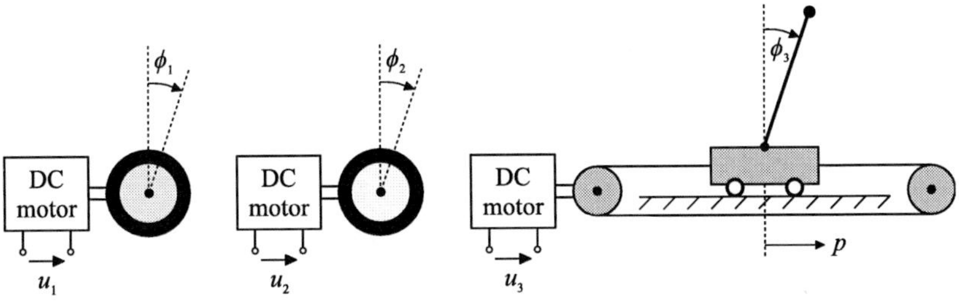

Figure 3.7: Simultaneous stabilization of two DC motors and one inverted pendulum

The state vector of the DC motor is defined as $\boldsymbol{x}_{\mathrm{p}i}(k) = \begin{pmatrix}\phi_i(t) & n_i(t)\end{pmatrix}^T$, where $\phi_i(t)$ is the angle of the wheel in radian and $n_i(t)$ is the speed of the wheel in rounds per second. The control input $u_i(t-\tau_i)$ is the armature voltage in volt. This leads to the continuous-time model (3.1) with the system and input matrices

$$\boldsymbol{A}_{\mathrm{p}i} = \begin{pmatrix} 0 & 6.2832 \\ 0 & -23.2303 \end{pmatrix}, \qquad \boldsymbol{B}_{\mathrm{p}i} = \begin{pmatrix} 0 \\ 5.1162 \end{pmatrix}$$

for $i \in \{1, 2\}$. The initial state vector is $\boldsymbol{x}_{\mathrm{p}i}(0) = \begin{pmatrix} 0 & 0 \end{pmatrix}^T$. The system and input matrices of the inverted pendulum are given by

$$\boldsymbol{A}_{\mathrm{p}3} = \begin{pmatrix} 0 & 1 & 0 & 0 \\ 30.0503 & 0 & 0 & 2.9156 \\ 0 & 0 & 0 & 1 \\ -0.5117 & 0 & 0 & -1.0015 \end{pmatrix}, \qquad \boldsymbol{B}_{\mathrm{p}3} = \begin{pmatrix} 0 \\ -2.4614 \\ 0 \\ 0.8455 \end{pmatrix}$$

with the state vector $\boldsymbol{x}_{\mathrm{c}3}(t) = \begin{pmatrix} \phi_3(t) & \dot{\phi}_3(t) & p(t) & \dot{p}(t) \end{pmatrix}$, where $\phi_3(t)$ is the angle in radian and $p(t)$ is the position in meter. The control input $u_3(t - \tau_3)$ is the armature voltage in volt of the actuating DC motor. For the inverted pendulum the initial state vector is unknown.

Based on a coordinate transformation that translates the reference state into the origin, the setpoint tracking problem can be considered as a regulation problem. The control and scheduling law is then applied on the deviation variables $\boldsymbol{x}_{\mathrm{a}i}(k) - \boldsymbol{x}_{\mathrm{a}i}^{\mathrm{ss}}$, which are defined based on the steady-state values $\boldsymbol{x}_{\mathrm{a}i}^{\mathrm{ss}}$, see [FPW98, Section 8.4] for a comprehensive presentation. Thus, the control and scheduling law is given by

$$\boldsymbol{u}(k) = \boldsymbol{K}_{j(k)} \left(\boldsymbol{x}(k) - \boldsymbol{x}^{\mathrm{ss}}\right) + \boldsymbol{u}^{\mathrm{ss}} \tag{3.81a}$$

$$j(k) = \arg\min_{l \in \mathbb{J}} \left(\boldsymbol{x}(k) - \boldsymbol{x}^{\mathrm{ss}}\right)^T \boldsymbol{P}_l \left(\boldsymbol{x}(k) - \boldsymbol{x}^{\mathrm{ss}}\right) \tag{3.81b}$$

where $\boldsymbol{x}^{\mathrm{ss}}$ and $\boldsymbol{u}^{\mathrm{ss}}$ are the steady state values of the holistic state and control input vector, defined in (3.12). In this experiment the steady state values are $\boldsymbol{x}_{\mathrm{a}i}^{\mathrm{ss}} = \begin{pmatrix} r_i & 0 & 0 \end{pmatrix}^T$ for $i \in \{1, 2\}$, $\boldsymbol{x}_{\mathrm{a}3}^{\mathrm{ss}} = \boldsymbol{0}$ and $\boldsymbol{u}^{\mathrm{ss}} = \boldsymbol{0}$.

For the offline scheduling approach the scheduling is realized following a periodic sequence. Based on the given sequence $(3, 2, 3, 1)$ the optimal control gain matrices are determined using periodic control [GIL07, BCD90, BCD91]. For both approaches the step size is set $h_i = \frac{1}{60}$ s for all $i \in \{1, 2, 3\}$. The input delay τ_i results from the execution times of the different steps, see the discussion in Chapter 2. Based on the measurement of the execution time of each step, see Table A.1, the input delay is given by

$$\tau_{1/2}^{\mathrm{on}} = 2.4946\,\mathrm{ms}, \quad \tau_{1/2}^{\mathrm{off}} = 0.3855\,\mathrm{ms},$$
$$\tau_{3}^{\mathrm{on}} = 2.5061\,\mathrm{ms}, \quad \tau_{3}^{\mathrm{off}} = 0.6209\,\mathrm{ms}.$$

For determining the control and scheduling parameters of both approaches the weighting matrices are selected as

$$\boldsymbol{Q}_{\mathrm{c}1/2} = \mathrm{diag}\left(15000, 150\right), \qquad \boldsymbol{R}_{\mathrm{c}1/2} = 3000,$$
$$\boldsymbol{Q}_{\mathrm{c}3} = \mathrm{diag}\left(8000, 200, 55000, 1\right), \quad \boldsymbol{R}_{\mathrm{c}3} = 4$$

after several design iterations. Thereby, the relation between the weighting matrices of the different plants is important for a good overall performance. This means, if for one plant the weighting is chosen too small compared with the other plants its performance may become unsatisfactory, as the weighting matrices affect the resulting scheduling matrices $\boldsymbol{P}_l$.

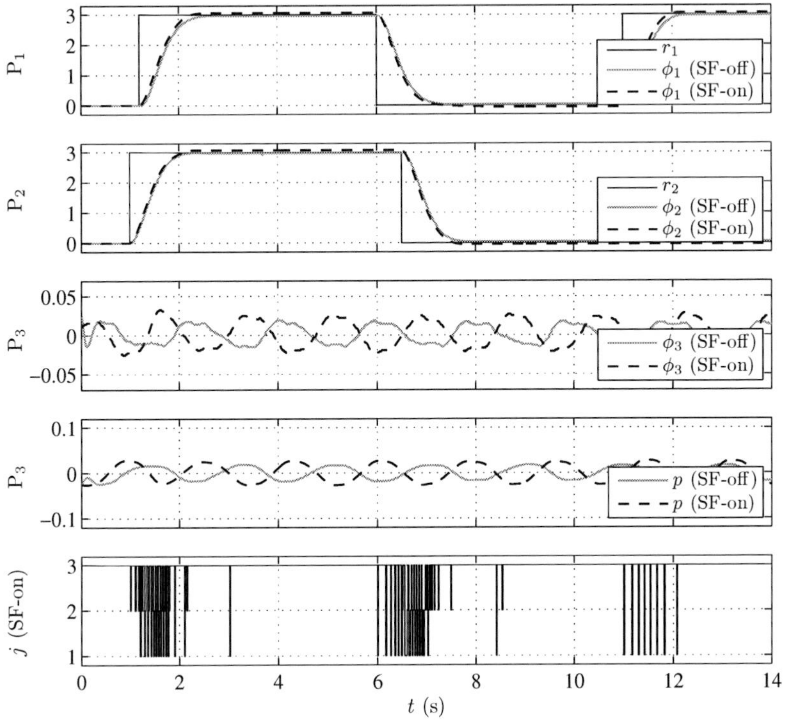

Figure 3.8: Experimental results under state-feedback control and scheduling

Both strategies are implemented on the microcontroller and the experiment is run for $T_{\text{eval}} = 14\,\text{s}$, which leads to the results given in Figure 3.8.

Under online scheduling most of the time the scheduling index is $j(k) = 3$, which corresponds to the inverted pendulum, as this plant requires continuous attention due to its instability. For the DC motors (plant P_1 and P_2) there is no necessity for an update of the control input if the plant is in the steady state. Therefore, the control signals u_1 and u_2 are only updated if there is a reference change and then until the steady state is reached. For plant P_1 the reference signal changes at $t = 1.2\,\text{s}$, $t = 6.0\,\text{s}$ and $t = 11.0\,\text{s}$ and for plant P_2 at $t = 1.0\,\text{s}$ and $t = 6.5\,\text{s}$. Considering the control performance the offline scheduling strategy shows a better performance for the inverted pendulum as the oscillation around zero is smaller both for the angle ϕ_3 and the position p. This is also shown by the resulting cost J_3 given in Table 3.2. For the control performance of the DC motors the online scheduling strategy shows its reactiveness to disturbances or reference

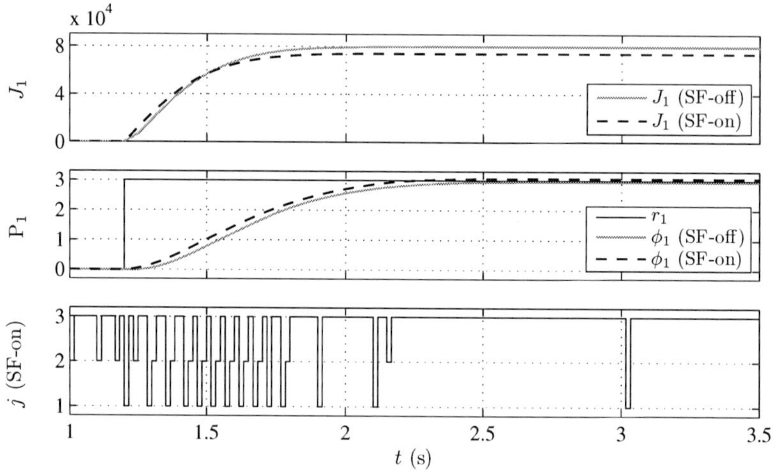

Figure 3.9: Experimental evaluation of plant P_1

changes, which is illustrated in Figure 3.9 for plant P_1. Under online scheduling the task index switches to $j(k) = 1$ at the time instant $k = 72$ ($t_k = 1.2\,\mathrm{s}$) of the reference change, which allows an immediate reaction. However, under offline scheduling the control input u_1 is only updated at the time instant $k = 75$ ($t_k = 1.25\,\mathrm{s}$), leading to a slower reaction and thus a worse control performance, see the cost J_1 in Figure 3.9. Under both scheduling approaches there is a small steady state error for plant P_1 and P_2, which is caused by a dead zone of the DC motor and is not related to the scheduling strategy. The resulting costs J_1 and J_2 given in Table 3.2 emphasize that the online scheduling strategy outperforms the offline scheduling approach considering the control performance of the plants P_1 and P_2. Finally, the online scheduling strategy can improve the overall control performance J compared with the offline scheduling. The experiment shows that online scheduling is especially beneficial for scheduling problems, where a reactiveness to disturbances or reference changes is required. On the other hand, offline scheduling shows its benefits for scheduling problems, where the plants need continuous attention.

Method	J_1	J_2	J_3	J
SF-off	$2.36 \cdot 10^5$	$1.53 \cdot 10^5$	$7.47 \cdot 10^2$	$3.90 \cdot 10^5$
SF-on	$2.33 \cdot 10^5$	$1.51 \cdot 10^5$	$9.41 \cdot 10^2$	$3.84 \cdot 10^5$

Table 3.2: Cost under different scheduling strategies

3.7.2 Two Electronic Double Integrator Circuits

In the second experimental study, two equivalent electronic double integrator circuits, see Figure 3.10, are considered to evaluate the output-feedback scheduling strategies (OF-PO-on, OF-CO-on) in comparison with the state-feedback online scheduling approach (SF-on). The continuous-time state equation is given by

$$\begin{aligned} \begin{pmatrix} \dot{v}_{1i}(t) \\ \dot{v}_{2i}(t) \end{pmatrix} &= \begin{pmatrix} 0 & \frac{-1}{R_1C_1} \\ 0 & 0 \end{pmatrix} \begin{pmatrix} v_{1i}(t) \\ v_{i2}(t) \end{pmatrix} + \begin{pmatrix} 0 \\ \frac{-1}{R_2C_2} \end{pmatrix} u_i(t-\tau_i) \\ y_i(t) &= \begin{pmatrix} 1 & 0 \end{pmatrix} \begin{pmatrix} v_{1i}(t) \\ v_{2i}(t) \end{pmatrix} \end{aligned} \tag{3.82}$$

see [LMVF12] for the plant details. The control aim is to make the output voltage v_{i1} track a constant reference signal r_i. This is realized in the same manner as in the previous experiment in Section 3.7.1 applying (3.81) with $\boldsymbol{x}_{\mathrm{a}i}^{\mathrm{ss}} = \begin{pmatrix} r_i & 0 & 0 \end{pmatrix}^T$ for $i \in \{1,2\}$ and $\boldsymbol{u}^{\mathrm{ss}} = \boldsymbol{0}$.

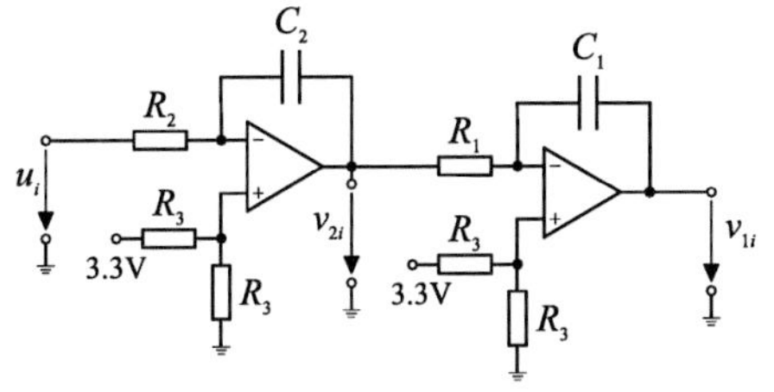

Figure 3.10: Electronic double integrator circuit

Component	Nominal value
R_3	$0.58\,\mathrm{k\Omega}$
R_1	$100\,\mathrm{k\Omega}$
R_2	$100\,\mathrm{k\Omega}$
C_1	$470\,\mathrm{nF}$
C_2	$470\,\mathrm{nF}$

Table 3.3: Parameters of electronic components

With the model parameters from Table 3.3, this results in the system matrices

$$\boldsymbol{A}_{\mathrm{p}i} = \begin{pmatrix} 0 & -21.2766 \\ 0 & 0 \end{pmatrix}, \quad \boldsymbol{B}_{\mathrm{p}i} = \begin{pmatrix} 0 \\ -21.2766 \end{pmatrix}, \quad \boldsymbol{C}_{\mathrm{p}i} = \begin{pmatrix} 1 & 0 \end{pmatrix}.$$

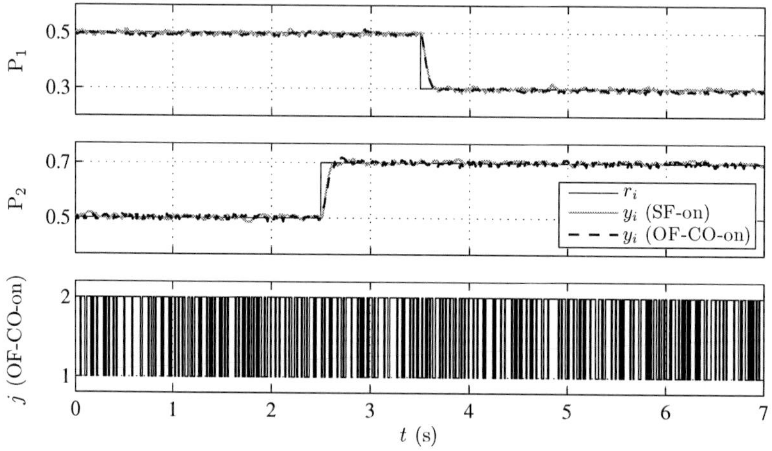

Figure 3.11: Experimental results

The design parameters for the control, scheduling and observer design are defined as

$$\boldsymbol{Q}_{ci} = \begin{pmatrix} 10 & 0 \\ 0 & 0 \end{pmatrix}, \quad \boldsymbol{R}_{ci} = 5, \quad \rho = 1$$

for all $i \in \{1, 2\}$. The proposed strategies are implemented on the microcontroller NXP LPC2294 with the step size $h_i = 10\,\text{ms}$ for $i \in \{1, 2\}$. The input delay varies according to the method with $\tau_i^{\text{SF-on}} = 0.841\,\text{ms}$, $\tau_i^{\text{OF-PO-on}} = 0.801\,\text{ms}$, and $\tau_i^{\text{OF-CO-on}} = 0.915\,\text{ms}$, $i \in \{1, 2\}$, which is obtained based on measurements given in Table A.2. The actuation is realized via pulse-width modulation (PWM).

In the given setup only the output voltage v_{1i} can be measured using A/D converters. Therefore, for realizing the state-feedback online scheduling approach (SF-on) the voltage v_{2i} is computed by differentiating the output signal, i.e. $v_{2i} = -R_1 C_1 \dot{v}_{1i}$, applying the backward difference approximation [ÅW90, Section 8.2]. For the output-feedback methods (OF-PO-on, OF-CO-on) the states are estimated applying the presented prediction and current observer, respectively.

The experimental results are given in Figure 3.11 for the methods SF-on and OF-CO-on. Both approaches show a similar output behavior. The scheduling index j is continuously switching between the two plants for both approaches, as can be seen in Figure 3.11 for the OF-CO-on strategy. The control signal u_2 is updated more frequently, which is due to the stronger measurement noise of plant P_2. The performance of the output-feedback online scheduling approach with prediction observer is very close to the performance with current observer and is therefore not depicted.

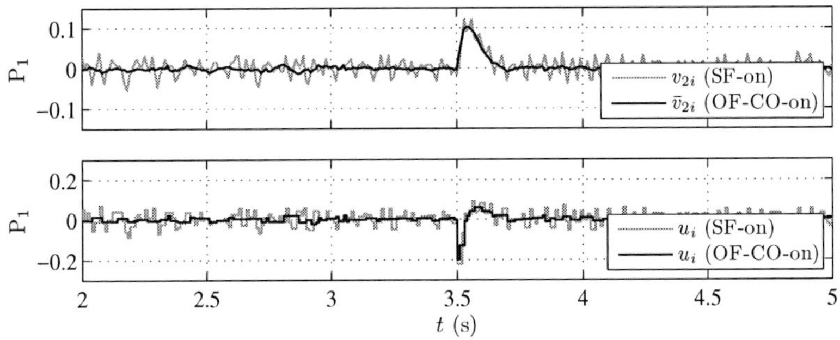

Figure 3.12: Experimental evaluation of plant P_1

Differences between the two shown approaches can be found in the resulting cost. As not all states can be measured, the cost is calculated based on the available state information, i.e. based on the measured output v_{1i} and the differentiated state v_{2i} for the state-feedback online scheduling approach and based on the estimated states $\begin{pmatrix}\bar{v}_{1i} & \bar{v}_{2i}\end{pmatrix}^T$ for the output-feedback online scheduling approaches. This results in $J_{\text{SF-on}} = 0.0997$, $J_{\text{OF-PO-on}} = 0.0611$ and $J_{\text{OF-CO-on}} = 0.0559$ showing that the output-feedback strategies outperform the state-feedback approach SF-on. The reason for this can be found in the state v_{2i} and the control input u_i, which are shown by means of plant P_1 in Figure 3.12. Obviously, the estimated state $\bar{v}_{2i}$ contains much less noise than the differentiated state v_{2i} leading to a smaller control input and consequently smaller costs. Prediction and current observer show very similar results for the state estimation, which also leads to similar costs.

In summary, the experiment shows that the proposed output-feedback online scheduling approach is a promising strategy if not all states are measurable.

3.8 Summary

In this chapter an output-based control and scheduling codesign strategy for scheduling the communication network is proposed. The output-based control and scheduling design problem is introduced and solved using the principle of separation. In the first step, a controller and scheduler are designed jointly assuming full state measurement. In the second step, the state estimation is realized applying a prediction and current observer, respectively. Due to the switched system nature, the validity of the separation principle needs to be verified. The effectiveness of the proposed scheduling methods is illustrated via simulative and experimental studies.

4 PI Control and Scheduling Design for Setpoint Tracking

The control and scheduling design approach introduced in the previous chapter showed a convincing performance, however, at the cost of a large scheduling overhead. Therefore, another output-based non-preemptive scheduling strategy in combination with output-feedback dynamic controllers is introduced, which especially focuses on a small scheduling overhead. Thereby, the focus lies on proportional-integral (PI) control for tracking a constant reference signal and/or rejecting constant disturbances.

4.1 Problem Formulation

Each plant P_i is described by the continuous-time state equation

$$\begin{aligned}\dot{\boldsymbol{x}}_{\mathrm{p}i}(t) &= \boldsymbol{A}_{\mathrm{p}i}\boldsymbol{x}_{\mathrm{p}i}(t) + \boldsymbol{B}_{\mathrm{p}i}\boldsymbol{u}_i(t-\tau_i) + \boldsymbol{B}_{\mathrm{w}i}\boldsymbol{w}_i(t)\\ \boldsymbol{y}_i(t) &= \boldsymbol{C}_{\mathrm{p}i}\boldsymbol{x}_{\mathrm{p}i}(t)\end{aligned} \tag{4.1}$$

where $\boldsymbol{A}_{\mathrm{p}i} \in \mathbb{R}^{n_i \times n_i}$ is the system matrix, $\boldsymbol{B}_{\mathrm{p}i} \in \mathbb{R}^{n_i \times m_i}$ is the input matrix, $\boldsymbol{B}_{\mathrm{w}i} \in \mathbb{R}^{n_i \times l_i}$ is the disturbance input matrix, $\boldsymbol{C}_{\mathrm{p}i} \in \mathbb{R}^{p_i \times n_i}$ is the output matrix, $\boldsymbol{x}_{\mathrm{p}i}(t) \in \mathbb{R}^{n_i}$ is the state vector, $\boldsymbol{y}_i(t) \in \mathbb{R}^{p_i}$ is the output vector, $\boldsymbol{u}_i(t-\tau_i) \in \mathbb{R}^{m_i}$ is control vector with the constant input delay τ_i defined in (2.1) and $\boldsymbol{w}_i(t) \in \mathbb{R}^{l_i}$ is the constant disturbance vector.

In this scheduling approach the time line is partitioned into time slots of a given length h, similar to [BÇH09]. At the time instant t_k, the beginning of each time slot, the outputs of all plants are sampled and fed to the scheduler. Based on this information a scheduling law is applied to determine the control task $\mathrm{T}_{j(k)}$ to be executed, with $j(k) \in \mathbb{J} = \{1, \ldots, M\}$. The computed control signal $\boldsymbol{u}_{j(k)}(k)$ is eventually forwarded to the plant $\mathrm{P}_{j(k)}$ at the time instant $t_k + \tau_{j(k)}$. For the other plants $\mathrm{P}_i \neq \mathrm{P}_{j(k)}$, the control signal is held constant until a new one is delivered. This represents a special case of the control task scheduling model introduced in Chapter 2, with $h_i = h$ for all $i \in \mathbb{J}$.

The step size $h = t_{k+1} - t_k$ is chosen larger than or equal to the maximum delay τ_i such that the execution of the scheduler and the selected control task can be finished within one time slot. The remaining idle time can be devoted to executing non-control tasks, see Figure 2.2. Additionally, by choosing the step size h appropriately a certain percentage of the processor resources can be reserved for non-control tasks.

Remark 4.1. An extension to a time-varying step size $h(k) = t_{k+1} - t_k = h_{j(k)}$ with respect to the task index $j(k)$, as introduced in Chapter 2, is also possible. For this purpose, the PI control structure and the stability analysis need to be adapted. This may be relevant if the execution times of the control tasks strongly differ and the processor resources shall be used only for control tasks, in case there are no non-control tasks. By selecting $h_{j(k)} = \tau_{j(k)}$ all processor resources are used for control tasks and there is no idle time.

In the following sections, the disturbance $\boldsymbol{w}_i(t)$ is neglected. In Section 4.5.3 the influence of the disturbance is then discussed.

Due to the discrete nature of the computation platform, a discrete-time representation of the continuous-time state equation (4.1) is crucial. Assuming the control signal of a plant P_i is updated within the interval $t_k \leq t < t_{k+1}$, i.e. $i = j(k)$, the control input signal is given as

$$\boldsymbol{u}_i(t-\tau_i) = \begin{cases} \boldsymbol{u}_i(t_{k-1}) & \text{for} \quad t_k \leq t < t_k + \tau_i \\ \boldsymbol{u}_i(t_k) & \text{for} \quad t_k + \tau_i \leq t < t_{k+1}. \end{cases} \tag{4.2}$$

Applying the discretization using zero-order hold on (4.1) results in the augmented discrete-time state equation

$$\begin{aligned} \boldsymbol{x}_{\mathrm{a}i}(k+1) &= \boldsymbol{A}_{\mathrm{a}i}(h)\boldsymbol{x}_{\mathrm{a}i}(k) + \boldsymbol{B}_{\mathrm{a}i}(h)\boldsymbol{u}_i(k) \\ \boldsymbol{y}_i(k) &= \boldsymbol{C}_{\mathrm{a}i}\boldsymbol{x}_{\mathrm{a}i}(k) \end{aligned} \tag{4.3}$$

with

$$\boldsymbol{x}_{\mathrm{a}i}(k) = \begin{pmatrix} \boldsymbol{x}_{\mathrm{p}i}(k)^T & \boldsymbol{u}_i(k-1) \end{pmatrix}^T \tag{4.4a}$$

$$\boldsymbol{A}_{\mathrm{a}i}(h) = \begin{pmatrix} e^{\boldsymbol{A}_{\mathrm{p}i}h} & \int_{h-\tau_i}^{h} e^{\boldsymbol{A}_{\mathrm{p}i}s} ds \boldsymbol{B}_{\mathrm{p}i} \\ \boldsymbol{0} & \boldsymbol{0} \end{pmatrix} \tag{4.4b}$$

$$\boldsymbol{B}_{\mathrm{a}i}(h) = \begin{pmatrix} \int_{0}^{h-\tau_i} e^{\boldsymbol{A}_{\mathrm{p}i}s} ds \boldsymbol{B}_{\mathrm{p}i} \\ \boldsymbol{I} \end{pmatrix} \tag{4.4c}$$

$$\boldsymbol{C}_{\mathrm{a}i} = \begin{pmatrix} \boldsymbol{C}_{\mathrm{p}i} & \boldsymbol{0} \end{pmatrix} \tag{4.4d}$$

see [ÅW90, Sec. 3.2] for details on the representation of systems with delayed inputs.

For the case $i \neq j(k)$ then $\boldsymbol{u}_i(t-\tau_i) = \boldsymbol{u}_i(t_{k-1})$ for $t_k \leq t < t_{k+1}$. The deviation from $\boldsymbol{u}_i(t_{k-1})$ to $\boldsymbol{u}_i(t_k)$ is considered in an additive error term in the modeling of the controller, see Section 4.2.

4.2 PI Control Structure

As the objective is to track a constant reference signal, a PI controller is applied, which is given by its digital implementation. For defining the PI controller, it is distinguished whether the control signal is updated or not within the interval $t_k \leq t < t_{k+1}$. If $i = j(k)$, then the integrator state and the control signal are updated, i.e.

$$\boldsymbol{x}_{\mathrm{I}i}(k) = \boldsymbol{x}_{\mathrm{I}i}(k-1) + \nu_i(k)\big(\boldsymbol{y}_i(k) - \boldsymbol{r}_i(k)\big) \tag{4.5a}$$

$$\boldsymbol{u}_i(k) = \boldsymbol{K}_{\mathrm{P}i}\big(\boldsymbol{y}_i(k) - \boldsymbol{r}_i(k)\big) + \boldsymbol{K}_{\mathrm{I}i}\boldsymbol{x}_{\mathrm{I}i}(k). \tag{4.5b}$$

If $i \neq j(k)$, then the integrator state and the control signal are not updated at all, i.e.

$$\boldsymbol{x}_{\mathrm{I}i}(k) = \boldsymbol{x}_{\mathrm{I}i}(k-1) \tag{4.6a}$$

$$\boldsymbol{u}_i(k) = \boldsymbol{u}_i(k-1), \tag{4.6b}$$

where $\boldsymbol{x}_{\mathrm{I}i}(k) \in \mathbb{R}^{p_i}$ is the integrator state, $\boldsymbol{K}_{\mathrm{P}i}, \boldsymbol{K}_{\mathrm{I}i} \in \mathbb{R}^{m_i \times p_i}$ are the proportional and integral gain, respectively. $\boldsymbol{r}_i(k) \in \mathbb{R}^{p_i}$ is the reference signal which is assumed to be constant and $\nu_i(k)$ indicates the integration time. The parameter $\nu_i(k)$ needs to be chosen appropriately with respect to the control update interval which is time-varying due to the scheduling.

There are different possibilities for defining $\nu_i(k)$, see also [VK06, DM09] for a detailed discussion. The intuitive choice for $\nu_i(k)$ would be the elapsed time since the previous control update $h_{\mathrm{act}_i}(k) = t_{k_i+1} - t_{k_i}$, i.e. the effective control update interval. Here t_{k_i} indicates the time instant, where the control input of the plant P_i is updated. However, the control update interval may become large due to the absence of a control update, for instance in the case that a plant is in the steady state. When the reference signal changes, the integral part explodes, which can result in large overshoot. Therefore, the control update interval is restricted, i.e. the parameter $\nu_i(k)$ is saturated to h_{max_i} when the control update interval exceeds the boundary h_{max_i}. Thus, $\nu_i(k)$ is defined as

$$\nu_i(k) = \begin{cases} h_{\mathrm{act}_i}(k) & \text{if} \quad h_{\mathrm{act}_i}(k) \leq h_{\mathrm{max}_i} \\ h_{\mathrm{max}_i} & \text{if} \quad h_{\mathrm{act}_i}(k) > h_{\mathrm{max}_i}. \end{cases} \tag{4.7}$$

Under the given assumption of a constant step size h the possible values of $\nu_i(k)$ are given by a set $\nu_i(k) = \ell h$ with $\ell \in \mathbb{H}_i = \{1, 2, ..., N_i\}$ where $h_{\mathrm{max}_i} = N_i \cdot h$.

Remark 4.2. The design parameter N_i needs to be selected taking into account the system dynamic of the plant P_i and the number of considered plants. On the one hand, the parameter N_i should be chosen not too large as a large integration time $\nu_i(k)$ can cause approximation errors for the discrete-time integration and thus a performance degradation. The procedure introduced in [VK06] proposes to limit the integration time $\nu_i(k)$ to the rise time T_{rise_i} of the plant P_i for a step response, i.e. $N_i \cdot h < T_{\mathrm{rise}_i}$. On the other hand, N_i should be selected not too small such that the integration time $\nu_i(k)$ equals the control update interval $h_{\mathrm{act}_i}(k)$ when the system is in the transient phase. In Section 4.3 the scheduler is designed in a way such that a plant in the transient phase is updated at least every M steps. Therefore, the variable N_i should be selected $N_i \geq M$. The influence of the integration time $\nu_i(k)$ also depends on the control parameter $\boldsymbol{K}_{\mathrm{I}i}$. A larger value of $\boldsymbol{K}_{\mathrm{I}i}$ will increase the influence of $\nu_i(k)$ on the control performance.

The logical variable

$$\delta_i(k) = \begin{cases} 1 & \text{if} \quad i = j(k) \\ 0 & \text{if} \quad i \neq j(k) \end{cases} \tag{4.8}$$

is introduced to model the general control law with respect to the task index $j(k)$, which is later used for modeling the closed-loop system.

In the following the error variable $\boldsymbol{e}_i(k)$ defined as

$$\begin{aligned} \boldsymbol{e}_i(k) &= \big(\boldsymbol{y}_i(k) - \boldsymbol{r}_i(k)\big) - \big(\hat{\boldsymbol{y}}_i(k-1) - \hat{\boldsymbol{r}}_i(k-1)\big) \\ &= \big(\boldsymbol{C}_{\mathrm{a}i}\boldsymbol{x}_{\mathrm{a}i}(k) - \boldsymbol{r}_i(k)\big) - \big(\boldsymbol{C}_{\mathrm{a}i}\hat{\boldsymbol{x}}_{\mathrm{a}i}(k-1) - \hat{\boldsymbol{r}}_i(k-1)\big) \end{aligned} \tag{4.9}$$

is introduced, where $\hat{\boldsymbol{y}}_i(k-1)$ and $\hat{\boldsymbol{r}}_i(k-1)$ indicate the output and reference, respectively, at the time instant of the most recent control update of the plant P_i before the time instant t_k and are defined as

$$\begin{pmatrix} \hat{\boldsymbol{y}}_i(k) \\ \hat{\boldsymbol{r}}_i(k) \end{pmatrix} = \begin{cases} \begin{pmatrix} \boldsymbol{y}_i(k) \\ \boldsymbol{r}_i(k) \end{pmatrix} & \text{if} \quad i = j(k) \\ \begin{pmatrix} \hat{\boldsymbol{y}}_i(k-1) \\ \hat{\boldsymbol{r}}_i(k-1) \end{pmatrix} & \text{if} \quad i \neq j(k). \end{cases} \tag{4.10}$$

This error gives an indication if the plant is in the transient state or in the steady state, based on which the necessity for a control update can be deduced. Assuming a constant reference signal, i.e. $\boldsymbol{r}_i(k) = \hat{\boldsymbol{r}}_i(k-1)$, a large error indicates the transient phase where a control update is necessary and a small error indicates that the plant is in the steady state. If the reference signal changes the error increases such that it can be concluded that an increasing error $\boldsymbol{e}_i(k)$ indicates the requirement for a control update, which is later used for the scheduling law in Section 4.3. This error is equivalent to the error used in [Årz99] for the event generator.

Based on the definition of the auxiliary variables $\hat{\boldsymbol{y}}_i(k)$ and $\hat{\boldsymbol{r}}_i(k)$ the control law (4.6) can be rewritten as

$$\boldsymbol{x}_{\mathrm{I}i}(k) = \boldsymbol{x}_{\mathrm{I}i}(k-1) \tag{4.11a}$$

$$\boldsymbol{u}_i(k) = \boldsymbol{K}_{\mathrm{P}i}\big(\hat{\boldsymbol{y}}_i(k-1) - \hat{\boldsymbol{r}}_i(k-1)\big) + \boldsymbol{K}_{\mathrm{I}i}\boldsymbol{x}_{\mathrm{I}i}(k) \tag{4.11b}$$

for the case that the control input is not updated, i.e. $i \neq j(k)$. By substituting (4.9) into (4.11b) the control law can then be rewritten with respect to the task index $j(k)$

$$\boldsymbol{x}_{\mathrm{I}i}(k) = \boldsymbol{x}_{\mathrm{I}i}(k-1) + \delta_i(k)\nu_i(k)\big(\boldsymbol{y}_i(k) - \boldsymbol{r}_i(k)\big) \tag{4.12a}$$

$$\boldsymbol{u}_i(k) = \boldsymbol{K}_{\mathrm{P}i}\Big(\boldsymbol{y}_i(k) - \boldsymbol{r}_i(k) - \big(1 - \delta_i(k)\big)\boldsymbol{e}_i(k)\Big) + \boldsymbol{K}_{\mathrm{I}i}\boldsymbol{x}_{\mathrm{I}i}(k). \tag{4.12b}$$

Finally, it is worth emphasizing that this PI controller is only applicable for systems which are stabilizable with an output-based PI controller.

4.3 Online Scheduler

For the design of the scheduler, the aim is to give a scheduler that allows to realize both a fast reaction to disturbances or reference changes and a low overhead. The scheduler is executed at the time instant t_k, the beginning of each time slot, in order to select one control task $\mathrm{T}_{j(k)}$ to be executed in the current time slot. The idea is to make a scheduling decision based on the error variable $\boldsymbol{e}_i(k)$ defined in (4.9) with a low scheduling overhead. Therefore, first a p-periodic sequence

$$\mathcal{S} := \big(s(0), s(1), ..., s(p-1)\big) \tag{4.13}$$

with $s(k) = s(k+p)$ and $s(k) \in \{1, 2, ..., M\}$ is introduced. The variable $s(k)$ indicates the plant $\mathrm{P}_i = \mathrm{P}_{s(k)}$ for which the event condition

$$\left\|\boldsymbol{e}_{s(k)}(k)\right\|_2 \geq \epsilon_{s(k)} \tag{4.14}$$

with $\epsilon_{s(k)} \in \mathbb{R}_0^+$ is verified in the current time slot, i.e. in each time slot only for one plant $\mathrm{P}_{s(k)}$ condition (4.14) is verified. If condition (4.14) is fulfilled for the considered plant $\mathrm{P}_i = \mathrm{P}_{s(k)}$, then the control task $\mathrm{T}_{s(k)}$ is executed, i.e. $j(k) = s(k)$. Otherwise, the plant $\mathrm{P}_{s(k)}$ can spare the resources and the scheduling index is determined based on

$$j(k) = \arg\max_{i \in \mathbb{J}} \lambda_i \left\|\boldsymbol{e}_i(k)\right\|_2 \tag{4.15}$$

with the design parameter $\lambda_i \in \mathbb{R}^+$. The parameter $\lambda_i \in \mathbb{R}^+$ allows a weighting of the norm of the error $\boldsymbol{e}_i(k)$, which is usually done according to the range of values of the output.

For simplicity, the periodic sequence is defined as

$$\mathcal{S} := \big(s(0), s(1), ..., s(M-1)\big) = \big(1, 2, ..., M\big) \tag{4.16}$$

with $s(k) = s(k+M)$ in the following. Thus, by selecting ϵ_i appropriately a plant in the transient phase will be updated at least every M steps. This motivates the lower bound $N_i \geq M$ for the integration time, see Remark 4.2. Based on the parameter ϵ_i a certain control performance can be guaranteed. By selecting $\epsilon_i = 0$, it can be guaranteed that every M steps the control task T_i is executed. If $\epsilon_i = 0$ is chosen for all plants, this leads to the static scheduling law $j(k) = (k \mod M) + 1$. However, for $\epsilon_i = 0$ the control task T_i would be also executed due to the presence of noise, even though there is no necessity for an update of the control signal. This may lead to an inefficient usage of the resources. Therefore, the threshold ϵ_i should be selected sufficiently large to avoid an execution of a control task due to noise. On the other hand, ϵ_i needs to be selected sufficiently small to guarantee a good performance, see also the discussion in Example 4.2 and in [Årz99] where condition (4.14) is applied in the context of event-triggered control.

The online scheduling procedure is formalized in Algorithm 4.1. The procedure can be also extended to other p-periodic sequences $\mathcal{S}$, which contain all plants.

Remark 4.3. The conditional branch in Algorithm 4.1 may lead to a varying execution time of the scheduler S. In the analysis only the worst-case execution time is considered.

Algorithm 4.1 Online scheduling

Input: k, $\boldsymbol{e}_i(k)$, ϵ_i, λ_i $\forall i$
Output: $j(k)$ // *scheduling index*
$s(k) = (k \mod M) + 1$
if $\left\|\boldsymbol{e}_{s(k)}(k)\right\|_2 \geq \epsilon_{s(k)}$ **then**
$j(k) = s(k)$
else
$j(k) = \arg\max_{i \in \mathbb{J}} \lambda_i \left\|\boldsymbol{e}_i(k)\right\|_2$
end if

4.4 Control Synthesis

After presenting the PI controller and scheduler, an open problem remains how to determine the control parameters. The control parameters are especially relevant for the transient control performance. If a plant is in the transient phase, it is expected that the control input is frequently updated. Further, the online scheduling is especially powerful in cases where not all control tasks need to be executed at the same time but when only one or a few control inputs need to be updated. Therefore, the PI controller needs to be aggressive especially under a small control update interval.

In general there are different methods to design the PI control parameters, see [ACL05] for an overview. In the following, the aim is to introduce a systematic design method which can guarantee a certain performance and is also applicable for MIMO systems. The majority of methods focuses on the design of continuous-time PI controllers. Systematic approaches to designing a PI controller with guaranteed cost are presented in [TPF97, XY04, RV07, MUTT09]. Most of those publications are only applicable for designing continuous-time PI controllers [TPF97, XY04, RV07]. Only [MUTT09] considers a discrete-time PI controller with the focus on a robust PI controller for an uncertain discrete-time system, assuming full state information. In order to design the parameters of the discrete-time PI controller, which is realized applying the backward integration and assumes only output information, a novel design approach is introduced. To realize an aggressive controller the control design is made under the assumption of a constant control update interval $h_{\mathrm{act}_i} = h$.

Remark 4.4. Also a robust control design with respect to a time-varying control update interval $h_{\mathrm{act}_i} = \ell h$, $\ell \in \mathbb{H}_i$ has been investigated. However, due to some resulting equality constraints in the LMI formulation this leads to very conservative results in case of a robust control design. Therefore, the control parameters are first designed assuming a constant control update interval $h_{\mathrm{act}_i} = h$ for each plant. Stability is analyzed a posteriori taking also the scheduling algorithm into account.

Under the constant control update interval h the PI control law (4.5) is given by

$$\boldsymbol{x}_{\mathrm{I}i}(k_i) = \boldsymbol{x}_{\mathrm{I}i}(k_i - 1) + h\big(\boldsymbol{y}_i(k_i) - \boldsymbol{r}_i(k_i)\big) \tag{4.17a}$$

$$\boldsymbol{u}_i(k_i) = \boldsymbol{K}_{\mathrm{P}i}\big(\boldsymbol{y}_i(k_i) - \boldsymbol{r}_i(k_i)\big) + \boldsymbol{K}_{\mathrm{I}i}\boldsymbol{x}_{\mathrm{I}i}(k_i). \tag{4.17b}$$

As the reference signal is constant for the following derivation, it can be set to zero without loss of generality. By combining (4.3) and (4.17) the overall closed-loop system is given by

$$\boldsymbol{z}_i(k_i + 1) = (\boldsymbol{A}_{\mathrm{z}i} + \boldsymbol{B}_{\mathrm{z}i}\boldsymbol{K}_i\boldsymbol{C}_{\mathrm{z}i})\,\boldsymbol{z}_i(k_i), \tag{4.18}$$

where

$$\boldsymbol{z}_i(k_i) = \begin{pmatrix}\boldsymbol{x}_{\mathrm{a}i}^T(k_i) & \boldsymbol{x}_{\mathrm{I}i}^T(k_i - 1)\end{pmatrix}^T \tag{4.19a}$$

$$\boldsymbol{K}_i = \begin{pmatrix}\boldsymbol{K}_{\mathrm{P}i} & \boldsymbol{K}_{\mathrm{I}i}\end{pmatrix} \tag{4.19b}$$

$$\boldsymbol{A}_{\mathrm{z}i} = \begin{pmatrix}\boldsymbol{A}_{\mathrm{a}i}(h) & \boldsymbol{0}\\ h\boldsymbol{C}_{\mathrm{a}i} & \boldsymbol{I}\end{pmatrix} \tag{4.19c}$$

$$\boldsymbol{B}_{\mathrm{z}i} = \begin{pmatrix}\boldsymbol{B}_{\mathrm{a}i}(h)\\ \boldsymbol{0}\end{pmatrix} \tag{4.19d}$$

$$\boldsymbol{C}_{\mathrm{z}i} = \begin{pmatrix}\boldsymbol{C}_{\mathrm{a}i} & \boldsymbol{0}\\ h\boldsymbol{C}_{\mathrm{a}i} & \boldsymbol{I}\end{pmatrix}. \tag{4.19e}$$

The PI control synthesis problem can now be formulated as

Problem 4.1 *For the plant* P_i, *find the control gain matrix* $\boldsymbol{K}_i = \begin{pmatrix}\boldsymbol{K}_{\mathrm{P}i} & \boldsymbol{K}_{\mathrm{I}i}\end{pmatrix}$ *such that the closed-loop system* (4.18) *is globally asymptotically stable (GAS) with guaranteed performance.*

The guaranteed performance is based on the continuous-time cost function

$$J_i = \int_0^\infty \begin{pmatrix}\boldsymbol{x}_{\mathrm{p}i}(t)\\ \boldsymbol{x}_{\mathrm{I}i}(t)\\ \boldsymbol{u}_i(t-\tau_i)\end{pmatrix}^T \begin{pmatrix}\boldsymbol{Q}_{\mathrm{c}i} & 0 & 0\\ 0 & \boldsymbol{Q}_{\mathrm{I}i} & 0\\ 0 & 0 & \boldsymbol{R}_{\mathrm{c}i}\end{pmatrix} \begin{pmatrix}\boldsymbol{x}_{\mathrm{p}i}(t)\\ \boldsymbol{x}_{\mathrm{I}i}(t)\\ \boldsymbol{u}_i(t-\tau_i)\end{pmatrix} dt \tag{4.20}$$

with $\boldsymbol{Q}_{\mathrm{c}i} \in \mathbb{R}^{n_i \times n_i}$, $\boldsymbol{Q}_{\mathrm{I}i} \in \mathbb{R}^{p_i \times p_i}$ symmetric and positive semidefinite and $\boldsymbol{R}_{\mathrm{c}i} \in \mathbb{R}^{m_i \times m_i}$ symmetric and positive definite. The discretized cost function using ZOH is given by

$$J_i = \sum_{k_i=0}^{\infty} \begin{pmatrix}\boldsymbol{z}_i(k_i)\\ \boldsymbol{u}_i(k_i)\end{pmatrix}^T \boldsymbol{Q}_i \begin{pmatrix}\boldsymbol{z}_i(k_i)\\ \boldsymbol{u}_i(k_i)\end{pmatrix}. \tag{4.21}$$

The discretization procedure for calculating the weighting matrix $\boldsymbol{Q}_i$ is explained in the Appendix A.1.2.

Substituting the PI control law (4.17b) $\boldsymbol{u}_i(k_i) = \boldsymbol{K}_i\boldsymbol{C}_{\mathrm{z}i}\boldsymbol{z}_i(k_i)$ into (4.21) yields the discrete-time closed-loop cost function

$$J_i = \sum_{k_i=0}^{\infty} \boldsymbol{z}_i^T(k_i)\tilde{\boldsymbol{Q}}_i\boldsymbol{z}_i(k_i) \tag{4.22}$$

where

$$\tilde{\boldsymbol{Q}}_i = \begin{pmatrix} \boldsymbol{I} \\ \boldsymbol{K}_i\boldsymbol{C}_{\mathrm{z}i} \end{pmatrix}^T \boldsymbol{Q}_i \begin{pmatrix} \boldsymbol{I} \\ \boldsymbol{K}_i\boldsymbol{C}_{\mathrm{z}i} \end{pmatrix}.$$

Consider the quadratic Lyapunov function

$$V_i(k_i) = \boldsymbol{z}_i^T(k_i)\boldsymbol{S}_i\boldsymbol{z}_i(k_i) \tag{4.23}$$

with the matrix $\boldsymbol{S}_i \in \mathbb{R}^{(n_i+m_i+p_i)\times(n_i+m_i+p_i)}$ symmetric and positive definite. For proving that the closed-loop system (4.18) is globally asymptotically stable (GAS), it needs to be shown that $\Delta V_i(k_i) = V_i(k_i+1) - V_i(k_i) < 0$ holds for all $\boldsymbol{z}_i(k_i) \neq \boldsymbol{0}$.

Theorem 4.1 *For a given control gain matrix $\boldsymbol{K}_i$, the discrete-time closed-loop system (4.18) is GAS, if there exist a matrix $\boldsymbol{G}_i \in \mathbb{R}^{(n_i+m_i+p_i)\times(n_i+m_i+p_i)}$ unrestricted and a matrix $\boldsymbol{Z}_i \in \mathbb{R}^{(n_i+m_i+p_i)\times(n_i+m_i+p_i)}$ symmetric and positive definite satisfying the LMI*

$$\begin{pmatrix} \boldsymbol{G}_i^T + \boldsymbol{G}_i - \boldsymbol{Z}_i & * & * \\ (\boldsymbol{A}_{\mathrm{z}i} + \boldsymbol{B}_{\mathrm{z}i}\boldsymbol{K}_i\boldsymbol{C}_{\mathrm{z}i})\boldsymbol{G}_i & \boldsymbol{Z}_i & * \\ \boldsymbol{Q}_i^{1/2}\begin{pmatrix} \boldsymbol{G}_i \\ \boldsymbol{K}_i\boldsymbol{C}_{\mathrm{z}i}\boldsymbol{G}_i \end{pmatrix} & 0 & \boldsymbol{I} \end{pmatrix} > 0. \tag{4.24}$$

Furthermore, the closed-loop cost function (4.22) *is bounded.*

Proof. Assume that (4.24) is satisfied. Then $\boldsymbol{G}_i^T + \boldsymbol{G}_i - \boldsymbol{Z}_i > \boldsymbol{0}$ and equivalently $\boldsymbol{G}_i^T + \boldsymbol{G}_i > \boldsymbol{Z}_i$ holds. Thus, $\boldsymbol{G}_i$ is of full rank and hence invertible. Since $\boldsymbol{Z}_i$ is positive definite, also

$$(\boldsymbol{Z}_i - \boldsymbol{G}_i)^T \boldsymbol{Z}_i^{-1}(\boldsymbol{Z}_i - \boldsymbol{G}_i) \geq \boldsymbol{0} \tag{4.25}$$

holds, which is equivalent to

$$\boldsymbol{G}_i^T\boldsymbol{Z}_i^{-1}\boldsymbol{G}_i \geq \boldsymbol{G}_i^T + \boldsymbol{G}_i - \boldsymbol{Z}_i. \tag{4.26}$$

Therefore, (4.24) implies

$$\begin{pmatrix} \boldsymbol{G}_i^T\boldsymbol{Z}_i^{-1}\boldsymbol{G}_i & * & * \\ (\boldsymbol{A}_{\mathrm{z}i} + \boldsymbol{B}_{\mathrm{z}i}\boldsymbol{K}_i\boldsymbol{C}_{\mathrm{z}i})\boldsymbol{G}_i & \boldsymbol{Z}_i & * \\ \boldsymbol{Q}_i^{1/2}\begin{pmatrix} \boldsymbol{G}_i \\ \boldsymbol{K}_i\boldsymbol{C}_{\mathrm{z}i}\boldsymbol{G}_i \end{pmatrix} & 0 & \boldsymbol{I} \end{pmatrix} > 0. \tag{4.27}$$

Substituting $\boldsymbol{Z}_i^{-1} = \boldsymbol{S}_i$ and pre-/post-multiplying (4.27) by $\mathrm{diag}(\boldsymbol{G}_i^{-T}, \boldsymbol{I}, \boldsymbol{I})$ and its transpose, respectively, results in

$$\begin{pmatrix} \boldsymbol{S}_i & * & * \\ \boldsymbol{A}_{\mathrm{z}i} + \boldsymbol{B}_{\mathrm{z}i}\boldsymbol{K}_i\boldsymbol{C}_{\mathrm{z}i} & \boldsymbol{S}_i^{-1} & * \\ \boldsymbol{Q}_i^{1/2}\begin{pmatrix} \boldsymbol{I} \\ \boldsymbol{K}_i\boldsymbol{C}_{\mathrm{z}i} \end{pmatrix} & 0 & \boldsymbol{I} \end{pmatrix} > 0. \tag{4.28}$$

Applying the Schur complement twice results in

$$\boldsymbol{S}_i - \tilde{\boldsymbol{A}}_{zi}^T \boldsymbol{S}_i \tilde{\boldsymbol{A}}_{zi} - \tilde{\boldsymbol{Q}}_i > 0 \tag{4.29}$$

with $\tilde{\boldsymbol{A}}_{zi} = \boldsymbol{A}_{zi} + \boldsymbol{B}_{zi}\boldsymbol{K}_i\boldsymbol{C}_{zi}$. Pre-/post-multiplying (4.29) with $\boldsymbol{z}_i^T(k_i)$ and its transpose, respectively, results in

$$\boldsymbol{z}_i^T(k_i)\left(\tilde{\boldsymbol{A}}_{zi}^T \boldsymbol{S}_i \tilde{\boldsymbol{A}}_{zi} - \boldsymbol{S}_i\right)\boldsymbol{z}_i(k_i) < -\boldsymbol{z}_i^T(k_i)\tilde{\boldsymbol{Q}}_i\boldsymbol{z}_i(k_i). \tag{4.30}$$

Therefore, (4.30) implies

$$\Delta V(k_i) < -\boldsymbol{z}_i^T(k_i)\tilde{\boldsymbol{Q}}_i\boldsymbol{z}_i(k_i). \tag{4.31}$$

Since the weighting matrix $\tilde{\boldsymbol{Q}}_i$ is positive semidefinite, the right-hand side of the inequality (4.31) can not be larger than zero, guaranteeing that (4.18) is GAS. Furthermore, summing up (4.31) over $k = 0, ..., \infty$ gives

$$V_i(0) - \lim_{k_i \to \infty} V_i(k_i) > J_i. \tag{4.32}$$

As the closed-loop system (4.18) is GAS, $V_i(k_i) \to 0$ as $k_i \to \infty$. Thus, inequality (4.32) implies an upper bound

$$J_i < \boldsymbol{z}_i^T(0)\boldsymbol{Z}_i^{-1}\boldsymbol{z}_i(0) < \mathrm{tr}(\boldsymbol{Z}_i^{-1})\|\boldsymbol{z}_i(0)\|_2^2 \tag{4.33}$$

on the cost function J_i. This completes the proof. □

In the next step a PI control synthesis is proposed for determining the control gain matrix $\boldsymbol{K}_i = \begin{pmatrix}\boldsymbol{K}_{\mathrm{P}i} & \boldsymbol{K}_{\mathrm{I}i}\end{pmatrix}$.

Theorem 4.2 *The solution to Problem 4.1 is obtained from the LMI optimization problem*

$$\min_{\boldsymbol{K}_i, \boldsymbol{Z}_i} \mathrm{tr}\left(\boldsymbol{Z}_i^{-1}\right) \quad \text{subject to} \tag{4.34a}$$

$$\boldsymbol{C}_{zi}\boldsymbol{G}_i = \boldsymbol{V}_i\boldsymbol{C}_{zi} \tag{4.34b}$$

$$\begin{pmatrix} \boldsymbol{G}_i^T + \boldsymbol{G}_i - \boldsymbol{Z}_i & * & * \\ \boldsymbol{A}_{zi}\boldsymbol{G}_i + \boldsymbol{B}_{zi}\boldsymbol{U}_i\boldsymbol{C}_{zi} & \boldsymbol{Z}_i & * \\ \boldsymbol{Q}_i^{1/2}\begin{pmatrix}\boldsymbol{G}_i \\ \boldsymbol{U}_i\boldsymbol{C}_{zi}\end{pmatrix} & 0 & \boldsymbol{I} \end{pmatrix} > 0 \tag{4.34c}$$

with the LMI variables $\boldsymbol{G}_i \in \mathbb{R}^{(n_i+m_i+p_i)\times(n_i+m_i+p_i)}$, $\boldsymbol{V}_i \in \mathbb{R}^{2p_i\times 2p_i}$ *and* $\boldsymbol{U}_i \in \mathbb{R}^{m_i\times 2p_i}$ *unrestricted, and* $\boldsymbol{Z}_i \in \mathbb{R}^{(n_i+m_i+p_i)\times(n_i+m_i+p_i)}$ *symmetric and positive definite. The control gain matrix results from*

$$\boldsymbol{K}_i = \boldsymbol{U}_i\boldsymbol{V}_i^{-1}. \tag{4.35}$$

Proof. Assume that the conditions (4.34b) and (4.34c) are satisfied and consider further that $\boldsymbol{C}_{zi}$ is of full-row rank, (4.34b) implies that $\boldsymbol{V}_i$ is of full rank and thus invertible. Combining (4.34b) and (4.35) results in

$$\boldsymbol{U}_i \boldsymbol{C}_{zi} = \boldsymbol{K}_i \boldsymbol{V}_i \boldsymbol{C}_{zi} = \boldsymbol{K}_i \boldsymbol{C}_{zi} \boldsymbol{G}_i, \tag{4.36}$$

see also [DRI02]. Substituting (4.36) into (4.34c) results in condition (4.24) of Theorem 4.1. Based on (4.34a) an upper bound on the cost function is minimized, which can be optimized without the knowledge of the initial state, see inequality (4.33). This completes the proof. □

Remark 4.5. For an output-based optimal control design the optimal control gain matrix depends on the initial state, see [LVS12, Section 8.1]. As the initial state is usually unknown, the aim is to design a controller for an average initial state. Therefore, the initial state is assumed to be a Gaussian random variable with mean $\mathbf{0}$ and covariance matrix $\boldsymbol{I}$. This results in

$$\mathrm{E}\left(\boldsymbol{z}_i^T(0)\boldsymbol{S}_i\boldsymbol{z}_i(0)\right) = \mathrm{tr}\left(\boldsymbol{S}_i\right), \tag{4.37}$$

where $\mathrm{E}(\cdot)$ denotes the expected value, which motivates the objective function in (4.34a), see also [ÅW90, page 338].

Remark 4.6. By applying the determined control parameters under the scheduling algorithm 4.1 the optimized upper bound (4.33) on the control performance cannot be guaranteed anymore. However, the evaluations show that the proposed procedure allows to determine suitable control parameters efficiently, see Section 4.6 and 4.7.

4.5 Stability Analysis

In order to derive a stability criterion which can guarantee stability of each plant, a closed-loop model is derived. Based on the closed-loop model a Lyapunov-based stability criterion is given taking the scheduling law according to Algorithm 4.1 into account. The stability analysis is conducted for each plant individually.

4.5.1 Closed-Loop System Model

In general, the stability can be proved by showing that there exists a Lyapunov function with a decreasing tendency [Bra98], but there is no necessity to analyze the Lyapunov function at each time instant t_k. Therefore, the value of the Lyapunov function of one plant P_i is only analyzed at the time instants the event condition is verified and the control signal is updated, respectively. Those instants are in the following indicated by t_{k_i} where $t_{k_i+1} = t_{k_i} + \beta h$. As the event condition is verified according to the M-periodic sequence (4.16), i.e. every M steps, also the Lyapunov function is analyzed at least every

M steps. As the control signal may be updated more frequently than every M steps, the possible values of β are given by the set $\beta \in \mathbb{I}_i = \{1, 2, ..., M\}$.

Remark 4.7. Under the M-periodic sequence (4.16), there is $\mathbb{I}_i = \mathbb{J}$ for all i. If a different p-periodic sequence is selected, the set $\mathbb{I}_i$ may be different for different values of i and will differ from $\mathbb{J}$, which is not considered in the following.

For simplicity the signal $\boldsymbol{r}_i(k_i)$ is considered to be zero. Based on (4.3) and (4.12) the discrete-time system is given by

$$\begin{aligned} \boldsymbol{x}_{\mathrm{a}i}(k_i+1) &= \boldsymbol{A}_{\mathrm{a}i}(\beta h)\boldsymbol{x}_{\mathrm{a}i}(k_i) + \boldsymbol{B}_{\mathrm{a}i}(\beta h)\boldsymbol{K}_{\mathrm{I}i}\big(x_{\mathrm{I}i}(k_i-1) + \alpha h \boldsymbol{C}_{\mathrm{a}i}\boldsymbol{x}_{\mathrm{a}i}(k_i)\big) \\ &\quad + \boldsymbol{B}_{\mathrm{a}i}(\beta h)\boldsymbol{K}_{\mathrm{P}i}\Big(\boldsymbol{C}_{\mathrm{a}i}\boldsymbol{x}_{\mathrm{a}i}(k_i) - \big(1-\delta_i(k_i)\big)\boldsymbol{e}_i(k_i)\Big) \end{aligned} \tag{4.38}$$

$$\boldsymbol{x}_{\mathrm{I}i}(k_i) = \boldsymbol{x}_{\mathrm{I}i}(k_i-1) + \alpha h \boldsymbol{C}_{\mathrm{a}i}\boldsymbol{x}_{\mathrm{a}i}(k_i) \tag{4.39}$$

with $\beta \in \mathbb{I}_i$ and $\alpha \in \mathbb{H}_i^0 = \{0, 1, 2, ..., N_i\}$, where N_i is a control design parameter as discussed in Remark 4.2. If $\alpha = 0$, this indicates there is no control update at the time instant t_{k_i}, i.e. $\|\boldsymbol{e}_i(k_i)\| \le \epsilon_i$, otherwise there is a control update and α gives an index for the value of $\nu_i(k_i) = \alpha h$. Further, the dynamic behavior of the error $\boldsymbol{e}_i(k_i)$ is analyzed. Based on (4.9) the error prediction $\boldsymbol{e}_i(k_i+1)$ is given by

$$\boldsymbol{e}_i(k_i+1) = \boldsymbol{C}_{\mathrm{a}i}\big(\boldsymbol{x}_{\mathrm{a}i}(k_i+1) - \boldsymbol{x}_{\mathrm{a}i}(k_i)\big) + \big(1-\delta_i(k_i)\big)\boldsymbol{e}_i(k_i). \tag{4.40}$$

Combining (4.38)-(4.40) and substituting $\delta_i(k_i) = \mathrm{sgn}(\alpha)$ the augmented closed-loop switched system of the plant P_i is given by

$$\begin{aligned} \boldsymbol{x}_i(k_i+1) &= \tilde{\boldsymbol{A}}_{i\alpha\beta}\boldsymbol{x}_i(k_i) \\ \boldsymbol{y}_i(k_i) &= \boldsymbol{C}_i\boldsymbol{x}_i(k_i) \end{aligned} \tag{4.41}$$

with $\boldsymbol{x}_i(k) = \big(\boldsymbol{x}_{\mathrm{a}i}^T(k) \quad \boldsymbol{x}_{\mathrm{I}i}^T(k-1) \quad \boldsymbol{e}_i^T(k)\big)^T$, $\boldsymbol{C}_i = \big(\boldsymbol{C}_{\mathrm{a}i} \quad \boldsymbol{0} \quad \boldsymbol{0}\big)$,

$$\tilde{\boldsymbol{A}}_{i\alpha\beta} = \begin{pmatrix} \boldsymbol{A}_{\mathrm{a}i}(\beta h)+\boldsymbol{B}_{\mathrm{a}i}(\beta h)(\boldsymbol{K}_{\mathrm{P}i}+\alpha h\boldsymbol{K}_{\mathrm{I}i})\boldsymbol{C}_{\mathrm{a}i} & \boldsymbol{B}_{\mathrm{a}i}(\beta h)\boldsymbol{K}_{\mathrm{I}i} & -(1-\mathrm{sgn}(\alpha))\boldsymbol{B}_{\mathrm{a}i}(\beta h)\boldsymbol{K}_{\mathrm{P}i} \\ \alpha h\boldsymbol{C}_{\mathrm{a}i} & \boldsymbol{I} & \boldsymbol{0} \\ \boldsymbol{C}_{\mathrm{a}i}\big(\boldsymbol{A}_{\mathrm{a}i}(\beta h)+\boldsymbol{B}_{\mathrm{a}i}(\beta h)(\boldsymbol{K}_{\mathrm{P}i}+\alpha h\boldsymbol{K}_{\mathrm{I}i})\boldsymbol{C}_{\mathrm{a}i}-\boldsymbol{I}\big) & \boldsymbol{C}_{\mathrm{a}i}\boldsymbol{B}_{\mathrm{a}i}(\beta h)\boldsymbol{K}_{\mathrm{I}i} & (1-\mathrm{sgn}(\alpha))\big(-\boldsymbol{C}_{\mathrm{a}i}\boldsymbol{B}_{\mathrm{a}i}(\beta h)\boldsymbol{K}_{\mathrm{P}i}+\boldsymbol{I}\big) \end{pmatrix}$$

and the switching indices $\beta \in \mathbb{I}_i$ and $\alpha \in \mathbb{H}_i^0$.

4.5.2 Stability criterion

Consider the switched Lyapunov function

$$W_i(k_i) = \boldsymbol{x}_i(k_i)^T \boldsymbol{P}_{i\delta_i(k_i)\gamma(k_i)}\boldsymbol{x}_i(k_i) \tag{4.42}$$

with $\boldsymbol{P}_{i\delta_i(k_i)\gamma(k_i)} \in \mathbb{R}^{(n_i+m_i+2p_i)\times(n_i+m_i+2p_i)}$ symmetric and positive definite and $\delta_i(k_i)$ as defined in (4.8). As the time interval between two observation of the Lyapunov function is varying, the index $\gamma(k_i)$ is introduced, indicating the instant of the preceding observation of the Lyapunov function, i.e. $t_{k_i-1} = t_{k_i} - \gamma(k_i)h$. As the Lyapunov function is analyzed at least every M steps the possible values of $\gamma(k_i)$ are given by the set $\gamma(k_i) \in \mathbb{I}_i = \{1, 2, ..., M\}$. The global asymptotic stability of the closed-loop system (4.41) can be shown if there exists a switched Lyapunov function (4.42) such that $W_i(k_i) > 0$ for all $\boldsymbol{x}_i \neq \boldsymbol{0}$ and $\Delta W_i(k) = W_i(k_i+1) - W_i(k_i) < 0$ along the trajectories of the closed-loop system for all $\boldsymbol{x}_i \neq \boldsymbol{0}$ under the scheduling Algorithm 4.1, see Theorem 1 in [FTCMM02]. The difference of the Lyapunov function is computed as

$$\Delta W_i(k_i) = \boldsymbol{x}_i^T(k_i)\left(\tilde{\boldsymbol{A}}_{i\alpha\beta}^T \boldsymbol{P}_{i\delta_i(k_i+1)\gamma(k_i+1)} \tilde{\boldsymbol{A}}_{i\alpha\beta} - \boldsymbol{P}_{i\delta_i(k_i)\gamma(k_i)}\right)\boldsymbol{x}_i(k_i). \tag{4.43}$$

Due to the absolute threshold in condition (4.14) it is difficult to give an asymptotic stability condition. Therefore, a practical stability criterion is derived, i.e. a Lyapunov function (4.42) needs to be found such that $\Delta W_i(k_i) < 0$ for all $\boldsymbol{x}_i \notin \mathcal{R}_i = \{\boldsymbol{x}_i : \boldsymbol{x}_i^T \boldsymbol{\Theta}_i \boldsymbol{x}_i \leq \eta_i\}$ with $\boldsymbol{\Theta}_i$ symmetric and positive definite and $\eta_i \in \mathbb{R}^+$. Thus, the convergence into the region $\mathcal{R}_i$ can be shown. The size of the region $\mathcal{R}_i$ can be designed by the choice of ϵ_i. For deriving a stability condition, the difference of the Lyapunov function (4.43) is analyzed for all possible pairs $\big(\delta_i(k_i), \delta_i(k_i+1)\big)$ considering the admissible values of α, β and γ as well as the event condition (4.14).

Case 1: For $\big(\delta_i(k_i), \delta_i(k_i+1)\big) = (1, 1)$ at both time instants t_{k_i} and t_{k_i+1} the control input is updated leading to the stability condition

$$\tilde{\boldsymbol{A}}_{i\alpha\beta}^T \boldsymbol{P}_{i\delta_i(k_i+1)\gamma(k_i+1)} \tilde{\boldsymbol{A}}_{i\alpha\beta} - \boldsymbol{P}_{i\delta_i(k_i)\gamma(k_i)} < \boldsymbol{0} \tag{4.44}$$

for all $\gamma(k_i+1) = \beta \in \mathbb{I}_i$ and $\gamma(k_i) \in \mathbb{I}_i$. The parameter α is either equal to $\gamma(k_i)$ or can be larger than $\gamma(k_i)$, if at the time instant of the preceding observation of the Lyapunov function the control input was not updated, i.e. $\alpha \in \{\gamma(k_i), ..., N_i\}$.

Case 2: For $\big(\delta_i(k_i), \delta_i(k_i+1)\big) = (1, 0)$ there is a control input update at the time instant t_{k_i} and at the time instant $t_{k_i+1} = t_{k_i} + \beta h$ condition (4.14) is verified but not satisfied such the control input is held. This leads to the stability condition (4.44) for all $\gamma(k_i+1) = \beta \in \mathbb{I}_i$, $\gamma(k_i) \in \mathbb{I}_i$ and $\alpha \in \{\gamma(k_i), ..., N_i\}$.

Case 3: For $\big(\delta_i(k_i), \delta_i(k_i+1)\big) = (0, 1)$ at the time instant t_{k_i} the condition $\|\boldsymbol{e}_i(k_i)\| < \epsilon_i$ is fulfilled such that the control input is not updated, i.e. $\alpha = 0$. At the time instant $t_{k_i+1} = t_{k_i} + \beta h$ with $\beta \in \mathbb{I}_i$ the control input is updated. In this case practical stability can be shown if

$$\boldsymbol{x}_i^T(k_i)\left(\tilde{\boldsymbol{A}}_{i\alpha\beta}^T \boldsymbol{P}_{i\delta_i(k_i+1)\gamma(k_i+1)} \tilde{\boldsymbol{A}}_{i\alpha\beta} - \boldsymbol{P}_{i\delta_i(k_i)\gamma(k_i)}\right)\boldsymbol{x}_i(k_i) < 0 \tag{4.45}$$

is satisfied for all $\boldsymbol{x}_i \notin \mathcal{R}_i$, $\beta \in \mathbb{I}_i$ and $\gamma(k_i) \in \mathbb{I}_i$ with $\alpha = 0$ and $\gamma(k_i+1) = \beta$.

Case 4: For $\big(\delta_i(k_i), \delta_i(k_i+1)\big) = (0, 0)$ at both time instants t_{k_i} and t_{k_i+1} the control input is not updated. Therefore, $\|\boldsymbol{e}_i(k_i)\| < \epsilon_i$ is fulfilled and $t_{k_i+1} = t_{k_i} + Mh$, as the event condition (4.14) is only verified every M steps. Thus, for proving stability condition (4.45) needs to be satisfied for all $\boldsymbol{x}_i \notin \mathcal{R}_i$ and $\gamma(k_i) \in \mathbb{I}_i$ with $\alpha = 0$ and $\beta = \gamma(k_i+1) = M$.

Based on the discussion the following theorem is given:

Theorem 4.3 *If there exist symmetric and positive definite matrices* $\boldsymbol{\Theta}_i$ *and* $\boldsymbol{P}_{i\delta_i\gamma}$, *with* $\delta_i \in \{0,1\}$, $\gamma \in \mathbb{I}_i$, *a positive scalar* $\eta_i \in \mathbb{R}^+$, *two scalars* $\mu_1, \mu_2 \in \mathbb{R}_0^+$ *and two a priori fixed scalars* $\kappa_1, \kappa_2 \in \mathbb{R}_0^+$ *such that the LMIs*

$$\tilde{\boldsymbol{A}}_{i\alpha\beta}^T \boldsymbol{P}_{i1\beta} \tilde{\boldsymbol{A}}_{i\alpha\beta} - \boldsymbol{P}_{i1\gamma} < \boldsymbol{0} \qquad \forall(\beta,\gamma,\alpha) \in \mathbb{H}_i \times \mathbb{H}_i \times \{\gamma, ..., N_i\} \tag{4.46a}$$

$$\tilde{\boldsymbol{A}}_{i\alpha\beta}^T \boldsymbol{P}_{i0\beta} \tilde{\boldsymbol{A}}_{i\alpha\beta} - \boldsymbol{P}_{i1\gamma} < \boldsymbol{0} \qquad \forall(\beta,\gamma,\alpha) \in \mathbb{H}_i \times \mathbb{H}_i \times \{\gamma, ..., N_i\} \tag{4.46b}$$

$$\boldsymbol{P}_{i0\gamma} - \tilde{\boldsymbol{A}}_{i\alpha\beta}^T \boldsymbol{P}_{i1\beta} \tilde{\boldsymbol{A}}_{i\alpha\beta} + \mu_1 \boldsymbol{E}_i - \kappa_1 \boldsymbol{\Theta}_i \geq \boldsymbol{0} \qquad \forall(\beta,\gamma) \in \mathbb{I}_i \times \mathbb{I}_i,\ \alpha = 0 \tag{4.46c}$$

$$-\mu_1 \epsilon_i^2 + \kappa_1 \eta_i > 0 \tag{4.46d}$$

$$\boldsymbol{P}_{i0\gamma} - \tilde{\boldsymbol{A}}_{i\alpha\beta}^T \boldsymbol{P}_{i0\beta} \tilde{\boldsymbol{A}}_{i\alpha\beta} + \mu_2 \boldsymbol{E}_i - \kappa_2 \boldsymbol{\Theta}_i \geq \boldsymbol{0} \qquad \forall \gamma \in \mathbb{I}_i,\ \beta = M,\ \alpha = 0 \tag{4.46e}$$

$$-\mu_2 \epsilon_i^2 + \kappa_2 \eta_i > 0 \tag{4.46f}$$

with $\boldsymbol{E}_i = \operatorname{diag}(\boldsymbol{0}, \boldsymbol{0}, \boldsymbol{I})$ *are feasible, then* $W_i(k_i) < 0$ *holds for all* $\boldsymbol{x}_i \notin \mathcal{R}_i$ *proving practical stability.*

Proof. If the LMI conditions (4.46a)-(4.46b) are feasible, then $\Delta W_i(k_i) < 0$ for all $\boldsymbol{x}_i \neq \boldsymbol{0}$ for $\delta_i(k_i) = 1$ as explained in *Case 1* and *Case 2*, see (4.43). Further, assume that the LMI conditions (4.46c)-(4.46d) are satisfied. Pre-/post-multiplying (4.46c) by the vector $\boldsymbol{x}_i^T(k_i)$ and its transpose, respectively, yields

$$\boldsymbol{x}_i^T(k_i) \left(\boldsymbol{P}_{i0\gamma} - \tilde{\boldsymbol{A}}_{i\alpha\beta}^T \boldsymbol{P}_{i1\beta} \tilde{\boldsymbol{A}}_{i\alpha\beta} + \mu_1 \boldsymbol{E}_i - \kappa_1 \boldsymbol{\Theta}_i \right) \boldsymbol{x}_i(k_i) \geq 0. \tag{4.47}$$

Summating (4.46d) and (4.47) leads to

$$\begin{aligned} &\boldsymbol{x}_i^T(k_i) \left(\tilde{\boldsymbol{A}}_{i\alpha\beta}^T \boldsymbol{P}_{i1\beta} \tilde{\boldsymbol{A}}_{i\alpha\beta} - \boldsymbol{P}_{i0\gamma} \right) \boldsymbol{x}_i(k_i) < \\ &- \mu_1 \left(\epsilon_i^2 - \boldsymbol{x}_i^T(k_i) \boldsymbol{E}_i \boldsymbol{x}_i(k_i) \right) - \kappa_1 \left(\boldsymbol{x}_i^T(k_i) \boldsymbol{\Theta}_i \boldsymbol{x}_i(k_i) - \eta_i \right) \end{aligned} \tag{4.48}$$

For $\delta_i(k_i) = 0$ the inequality $\epsilon_i^2 - \boldsymbol{x}_i^T(k_i) \boldsymbol{E}_i \boldsymbol{x}_i(k_i) > 0$ is satisfied, as $\|\boldsymbol{e}_i(k_i)\| < \epsilon_i$ holds. Further, if $\boldsymbol{x}_i \notin \mathcal{R}_i$ then $\boldsymbol{x}_i^T(k_i) \boldsymbol{\Theta}_i \boldsymbol{x}_i(k_i) - \eta_i > 0$ holds. Therefore, the right-hand side of (4.48) is always smaller than or equal to zero. Thus, $\Delta W_i(k_i) < 0$ for all $\boldsymbol{x}_i \notin \mathcal{R}_i$ for *Case 3*. In an analog way based on (4.46e)-(4.46f) it can be shown that $\Delta W_i(k_i) < 0$ holds for all $\boldsymbol{x}_i \notin \mathcal{R}_i$ for *Case 4*. Thus, $\Delta W_i(k_i) < 0$ is satisfied for all $\boldsymbol{x}_i \notin \mathcal{R}_i$, which completes the proof. □

Remark 4.8. One way to determine scalars $\kappa_1, \kappa_2 \in \mathbb{R}_0^+$ is to grid them up. As (4.46) is a feasibility and not an optimization problem the gridding of the unknown scalars can be made quite sparsely [Pet03].

Remark 4.9. By selecting $\epsilon_i = 0$ for a plant P_i, condition (4.14) is always fulfilled for this plant, i.e. the control input is updated at least every M steps. Therefore, only condition (4.46a) needs to be considered in the stability analysis. The other LMIs (4.46b)-(4.46f) can be omitted, as they comprise the *Cases 2-4*, in which condition (4.14) is not satisfied at the time instant t_{k_i} and/or t_{k_i+1}. This leads to a less conservative stability condition, which can prove asymptotic stability. As a consequence of this, the plant P_i cannot spare resources to the benefit of other plants anymore, which can lead to an inefficient resource utilization, as discussed in Section 4.3.

4.5.3 References and Disturbances

In case of a non-zero constant reference signal $\boldsymbol{r}_i(k) \neq \mathbf{0}$ and constant disturbances the stability proof is still valid, as the integral part of the controller allows to track the reference signal and reject the disturbance, respectively. However, the equilibrium point is not the origin. Therefore, a coordinate transformation is required to shift the equilibrium point into the origin. The Lyapunov function is then defined with respect to the state deviation variables, which are defined as

$$\begin{pmatrix} \tilde{\boldsymbol{x}}_{\mathrm{p}i}(k) \\ \tilde{\boldsymbol{u}}_i(k-1) \\ \tilde{\boldsymbol{x}}_{\mathrm{I}i}(k-1) \\ \tilde{\boldsymbol{e}}_i(k) \end{pmatrix} = \begin{pmatrix} \boldsymbol{x}_{\mathrm{p}i}(k) - \boldsymbol{x}_{\mathrm{p}i}^{\mathrm{ss}} \\ \boldsymbol{u}_i(k-1) - \boldsymbol{u}_i^{\mathrm{ss}} \\ \boldsymbol{x}_{\mathrm{I}i}(k-1) - \boldsymbol{x}_{\mathrm{I}i}^{\mathrm{ss}} \\ \boldsymbol{e}_i(k) - \boldsymbol{e}_i^{\mathrm{ss}} \end{pmatrix} \tag{4.49}$$

with $\boldsymbol{x}_{\mathrm{p}i}^{\mathrm{ss}}$, $\boldsymbol{x}_{\mathrm{I}i}^{\mathrm{ss}}$, $\boldsymbol{u}_i^{\mathrm{ss}}$ and $\boldsymbol{e}_i^{\mathrm{ss}}$ denoting the steady-state values, which are given by

$$\begin{pmatrix} \boldsymbol{x}_{\mathrm{p}i}^{\mathrm{ss}} \\ \boldsymbol{x}_{\mathrm{I}i}^{\mathrm{ss}} \end{pmatrix} = \begin{pmatrix} \boldsymbol{A}_{\mathrm{p}i} & \boldsymbol{B}_{\mathrm{p}i}\boldsymbol{K}_{\mathrm{I}i} \\ \boldsymbol{C}_{\mathrm{p}i} & 0 \end{pmatrix}^{-1} \begin{pmatrix} -\boldsymbol{B}_{\mathrm{w}i}\boldsymbol{w}_i \\ \boldsymbol{r}_i \end{pmatrix} \tag{4.50a}$$

$$\boldsymbol{u}_i^{\mathrm{ss}} = \boldsymbol{K}_{\mathrm{I}i}\boldsymbol{x}_{\mathrm{I}i}^{\mathrm{ss}} \tag{4.50b}$$

$$\boldsymbol{e}_i^{\mathrm{ss}} = \mathbf{0}. \tag{4.50c}$$

The steady-state values (4.50) are derived from (4.1) and (4.5b) setting $\dot{\boldsymbol{x}}_{\mathrm{p}i}(t) = \mathbf{0}$ and $\boldsymbol{y}_i(t) = \boldsymbol{r}_i(t)$. It is worth to note that the control synthesis and the stability analysis can be studied without knowing the steady-state values. They are only relevant for evaluating the cost function and the Lyapunov function in an application.

4.6 Illustrative Example and Comparison

For evaluating the effectiveness of the proposed scheduling approach, a comparison of the presented online scheduling method with a non-preemptive offline scheduling approach and the preemptive earliest-deadline first (EDF) scheduling algorithm is made based on simulation using the MATLAB toolbox TrueTime [OHC07, CHL+03].

4.6.1 Related Scheduling Approaches

For comparison the set of plants is controlled under a simple offline scheduling approach, i.e. the control tasks are scheduled non-preemptively under the M-periodic sequence

$$\mathcal{S} := \big(j(0), j(1), ..., j(M-1)\big) = \big(1, 2, ..., M\big) \tag{4.51}$$

with $j(k) = j(k+M)$. This results in a constant control update interval $h_{\mathrm{act}_i} = M \cdot h^{\mathrm{off}}$ for each plant P_i, where $h^{\mathrm{off}} = t_{k+1} - t_k$ indicates the step size under offline scheduling, which may differ from the one used for online scheduling.

Further, the earliest-deadline first (EDF) scheduling algorithm [LL73] is applied for controlling the set of plants. Therefore, a period T_i needs to be fixed for each control task, which is set equal to the deadline $D_i = T_i$. The period corresponds to the control update interval, i.e. $h_{\mathrm{act}_i} = T_i$, which may be different for each control task.

For both comparative approaches, the PI controller is realized applying the forward integration, i.e.

$$\boldsymbol{x}_{\mathrm{I}i}(k_i+1) = \boldsymbol{x}_{\mathrm{I}i}(k_i) + h_{\mathrm{act}_i}\big(\boldsymbol{y}_i(k_i) - \boldsymbol{r}_i(k_i)\big) \tag{4.52a}$$

$$\boldsymbol{u}_i(k_i) = \boldsymbol{K}_{\mathrm{P}i}^{\mathrm{EDF/off}}\big((\boldsymbol{y}_i(k_i) - \boldsymbol{r}_i(k_i)\big) + \boldsymbol{K}_{\mathrm{I}i}^{\mathrm{EDF/off}}\boldsymbol{x}_{\mathrm{I}i}(k_i). \tag{4.52b}$$

The control parameters are determined in a similar way as in Theorem 4.2, however taking into account the implementation of the PI controller using the forward integration, see Appendix A.3 for details.

4.6.2 Evaluation

For evaluating the different event-triggered PI control strategies two measures are taken into account, considering the control performance and the resource utilization. The resource utilization is defined as a percentage of the use of the processor

$$U = \left(\frac{1}{T_{\mathrm{sim}}}\sum_{i=1}^{M} E_i\right) \cdot 100, \tag{4.53}$$

where E_i is the accumulated processor time of a control task T_i during the simulation time T_{sim}, see also [LMV+13]. Under the EDF scheduling algorithm, the utilization can equivalently be computed by

$$U = \sum_{i=1}^{M} \frac{C_i}{T_i} \cdot 100, \tag{4.54}$$

where C_i is the worst-case execution time of the control task T_i [LL73]. The control performance is measured by the evaluation cost function

$$J_{\mathrm{eval}} = \sum_{i=1}^{M} J_{\mathrm{eval}i} = \sum_{i=1}^{M} \int_0^{T_{\mathrm{sim}}} \begin{pmatrix} \tilde{\boldsymbol{x}}_{\mathrm{p}i}(t) \\ \tilde{\boldsymbol{u}}_i(t-\tau_i) \end{pmatrix}^T \begin{pmatrix} \boldsymbol{Q}_{\mathrm{c}i} & \boldsymbol{0} \\ \boldsymbol{0} & \boldsymbol{R}_{\mathrm{c}i} \end{pmatrix} \begin{pmatrix} \tilde{\boldsymbol{x}}_{\mathrm{p}i}(t) \\ \tilde{\boldsymbol{u}}_i(t-\tau_i) \end{pmatrix} dt. \tag{4.55}$$

This corresponds to the sum of the cost (4.20) with $\boldsymbol{Q}_{\mathrm{I}i} = \mathbf{0}$. There are two reason for neglecting the integrator state $\boldsymbol{x}_{\mathrm{I}i}(k)$ in the evaluation cost function $J_{\mathrm{eval}i}$. First, the state vector $\boldsymbol{x}_{\mathrm{I}i}(k)$ does not give a real indication on the control performance. Second, the steady state $\boldsymbol{x}_{\mathrm{I}i}^{\mathrm{ss}}$ depends on the control parameter $\boldsymbol{K}_{\mathrm{I}i}$. Therefore, the initial states of $\tilde{\boldsymbol{x}}_{\mathrm{I}i}(t)$ may differ under different scheduling algorithms, as the algorithms are realized with different control parameters. Consequently, the cost J_i is not considered as a suitable performance measure, see also the discussion in Section 5.4.2.

Example 4.1 Consider a set of four plants, which are controlled using one embedded processor. The system matrices are given in Table 4.1. The plant P_1 describes the dynamic behavior of a satellite for the attitude control [DB05, Chapter 5] and the remaining plants P_i, $i \in \{2, 3, 4\}$ describe the dynamic behavior of a DC motor. For the plants P_2 and P_3 the system matrices are given for the control of the angular velocity and for the plant P_4 the control aim is to control the angular position to a constant reference, i.e. the state vector represents the angular position and the angular velocity.

Plant	P_1	$\mathrm{P}_{2/3}$	P_4
$\boldsymbol{A}_{\mathrm{p}i}$	$\begin{pmatrix} -1 & 1 \\ 0 & -9 \end{pmatrix}$	-10	$\begin{pmatrix} 0 & 1 \\ 0 & -10 \end{pmatrix}$
$\boldsymbol{B}_{\mathrm{p}i}$ / $\boldsymbol{B}_{\mathrm{w}i}$	$\begin{pmatrix} 0 \\ 4 \end{pmatrix}$	5	$\begin{pmatrix} 0 \\ 5 \end{pmatrix}$
$\boldsymbol{C}_{\mathrm{p}i}$	$\begin{pmatrix} 2.5 & 0 \end{pmatrix}$	1	$\begin{pmatrix} 1 & 0 \end{pmatrix}$
$\boldsymbol{Q}_{\mathrm{c}i}$	$\begin{pmatrix} 80 & 0 \\ 0 & 0 \end{pmatrix}$	20	$\begin{pmatrix} 12 & 0 \\ 0 & 0 \end{pmatrix}$
$\boldsymbol{Q}_{\mathrm{I}i}$	18	300	10^{-3}
$\boldsymbol{R}_{\mathrm{c}i}$	1.2	5	0.15

Table 4.1: System and weighting matrices for example 4.1

Under the proposed online scheduling the input delay is $\tau_i = 8\,\mathrm{ms}$, which corresponds to worst-case execution time, and the step size is set $h = 30\,\mathrm{ms}$. Under offline scheduling the input delay is $\tau_i^{\mathrm{off}} = 4\,\mathrm{ms}$ for all $i \in \mathbb{J}$ and the step size is selected as $h^{\mathrm{off}} = 15\,\mathrm{ms}$. Thus, $h_{\mathrm{act}_i} = 60\,\mathrm{ms}$ for all $i \in \mathbb{J}$ for the realization of the PI controller (4.52). For the EDF scheduling algorithm, the periods are defined as $T_1 = 70\,\mathrm{ms}$, $T_2 = T_3 = 60\,\mathrm{ms}$ and $T_4 = 50\,\mathrm{ms}$. The execution time C_i is equal to the input delay τ_i^{off} for all $i \in \mathbb{J}$. In the case of EDF scheduling, $h_{\mathrm{act}_i} = T_i$ for all $i \in \mathbb{J}$ for the realization of the PI controller (4.52). Based on the selected parameters, all approaches work with a similar resource utilization U, i.e. $U^{\mathrm{on}} = U^{\mathrm{off}} = 26.7\,\%$ and $U^{\mathrm{EDF}} = 27.0\,\%$, such that the scheduling overhead of the online scheduler is taken into account.

The control parameters $K_{\mathrm{P}i}$ and $K_{\mathrm{I}i}$ are designed with the weighting matrices given in

Table 4.1. For the online scheduling Theorem 4.2 is applied, which leads to

$$\boldsymbol{K}_1 = \begin{pmatrix}-2.2450 & -2.4911\end{pmatrix}, \boldsymbol{K}_{2/3} = \begin{pmatrix}-1.0183 & -6.8737\end{pmatrix}, \boldsymbol{K}_4 = \begin{pmatrix}-5.6932 & -0.0508\end{pmatrix}.$$

Under offline scheduling and EDF scheduling the control parameters are determined applying Theorem A.1, which leads to

$$\begin{aligned}
\boldsymbol{K}_1^{\text{off}} &= \begin{pmatrix}-2.4543 & -2.5185\end{pmatrix}, & \boldsymbol{K}_1^{\text{EDF}} &= \begin{pmatrix}-2.4621 & -2.4965\end{pmatrix},\\
\boldsymbol{K}_{2/3}^{\text{off}} &= \begin{pmatrix}-1.2193 & -6.4631\end{pmatrix}, & \boldsymbol{K}_{2/3}^{\text{EDF}} &= \begin{pmatrix}-1.2193 & -6.4631\end{pmatrix},\\
\boldsymbol{K}_4^{\text{off}} &= \begin{pmatrix}-5.8304 & -0.0520\end{pmatrix}, & \boldsymbol{K}_4^{\text{EDF}} &= \begin{pmatrix}-5.8575 & -0.0523\end{pmatrix}.
\end{aligned}$$

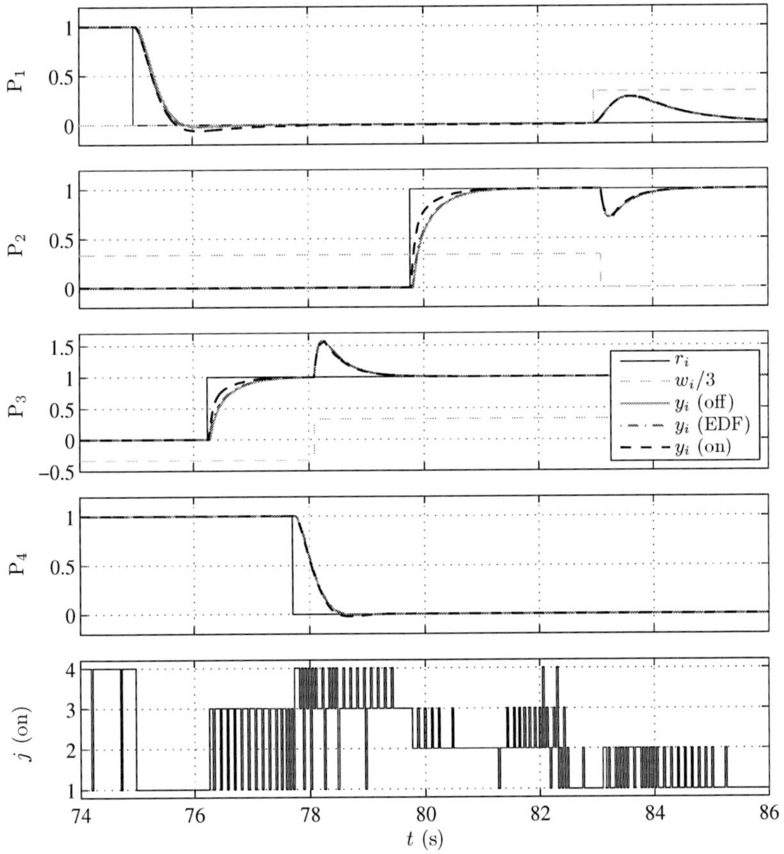

Figure 4.1: Simulation results

For the control synthesis the LMI optimization problem is implemented in MATLAB using the toolbox CVX [GB12, GB08] and the SeDuMi solver [Stu99]. The remaining control and scheduling parameters for the online scheduler are selected as $\lambda_i = 1$, $\epsilon_i = 0.001$ and $N_i = 4$ for all $i \in \mathbb{J}$. After specifying the control and scheduling parameters, stability is verified successfully for all plants by applying Theorem 4.3. The LMI feasibility problem is solved in MATLAB using the toolbox YALMIP [Löf04] and the SeDuMi solver [Stu99].

A simulation with a simulation time $T_{\text{sim}} = 100\,\text{s}$ is realized with piecewise constant reference signals and disturbances. The reference signals are created randomly, i.e. the reference switches between $r_i = 0$ and $r_i = 1$, where the duration between two consecutive reference changes is uniformly distributed in the interval $[10\,\text{s}, 20\,\text{s}]$. For the plants P_1, P_2 and P_3 also a piecewise constant disturbance signal is added, which switches between $w_i = -1$, $w_i = 0$ and $w_i = 1$. The duration between two consecutive changes of the disturbance signal is uniformly distributed in the interval $[20\,\text{s}, 30\,\text{s}]$. The initial states are $\boldsymbol{x}_{\text{p}i}(0) = \mathbf{0}$, $\hat{\boldsymbol{y}}_i(0) = 0$ and $\hat{\boldsymbol{r}}_i(0) = 0$ for all $i \in \mathbb{J}_i$. An excerpt of the output behavior, the reference signal and the disturbance signal is shown in Figure 4.1 for all plants. The scheduling index $j(k)$, also given in Figure 4.1, shows how the scheduler reacts to the reference and disturbance changes. Especially for plant P_2 and P_3, Figure 4.1 shows that the online scheduler reacts better to the reference change leading to a faster convergence. Additionally, the deviation due to the changing disturbance signal is smaller under online scheduling. The offline and EDF scheduler show a similar behavior. This is also reflected by the cost $J_{\text{eval}i}$, which is given for each plant under the different scheduling strategies in Table 4.2. For all plants the proposed online scheduling leads to the smallest cost.

Method	J_{eval1}	J_{eval2}	J_{eval3}	J_{eval4}	J_{eval}
offline scheduling	34.17	37.32	39.06	29.80	140.35
EDF scheduling	35.38	40.11	35.19	29.70	140.38
online scheduling	33.65	26.06	25.39	28.60	113.70

Table 4.2: Cost under different scheduling strategies

Example 4.2 In a second simulative study, the influence of the design parameters ϵ_i and λ_i is analyzed for proposed online scheduling strategy with the same four plants and the same corresponding control parameters $\boldsymbol{K}_i$ and N_i as in Example 4.1. For the plants P_2, P_3 and P_4 the scheduling parameters are set $\lambda_i = 1$ and $\epsilon_i = 0.001$ with $i \in \{2, 3, 4\}$, whereas for the plant P_1 different values of ϵ_i and λ_i are applied. The resulting control performance for the different plants is summarized in Figure 4.2 and Table 4.3. Figure 4.2 gives the cost J_{eval1} and J_{eval2} with respect to different values of ϵ_1 and λ_1. The reference and disturbance signals are equal to the ones used in Example 4.1. The first plot, which shows the cost J_{eval1}, emphasizes that by selecting $\epsilon_1 \leq 0.004$ small enough a certain control performance with $J_{\text{eval1}} \leq 33.8$ can be achieved. In this range, the parameter λ_1 has only minor influence on the control performance. However

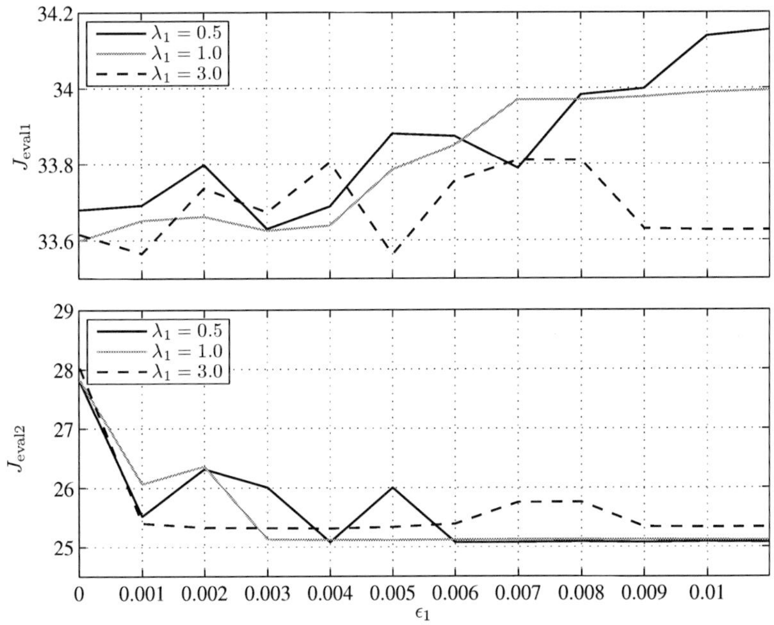

Figure 4.2: Evaluation of the control performance for different values of ϵ_1 and λ_1

for $\epsilon_1 > 0.004$, the control performance of the plant P_1 essentially depends on λ_1. For $\lambda_1 = 0.5$ and $\lambda_1 = 1.0$ the control performance degrades with increasing ϵ_1, whereas for $\lambda_1 = 3.0$ the control performance remains similar. On the other hand, the parameter ϵ_1 also affects the control performance of the other plants P_i, $i \in \{2, 3, 4\}$. This is shown in the second plot in Figure 4.2 exemplarily for plant P_2. For small values of ϵ_1 less resources can be spared, which makes the scheduler less flexible for reacting on disturbances or reference changes of the other plants. Therefore, the control performance of the plant P_2 generally improves with increasing ϵ_1. Similar effects can also be seen

Method	J_{eval1}	J_{eval2}	J_{eval3}	J_{eval4}
online scheduling ($\epsilon_1 = 2 \cdot 10^{-5}$)	33.60	27.82	24.82	28.95
online scheduling ($\epsilon_1 = 2 \cdot 10^{-3}$)	33.66	26.36	25.27	28.59
online scheduling ($\epsilon_1 = 6 \cdot 10^{-3}$)	33.85	25.12	23.67	28.58

Table 4.3: Cost for different values of ϵ_1 with $\lambda_1 = 1$

for the plants P_3 and P_4 as shown in Table 4.3. Additionally, for higher values of ϵ_1 the scheduling is more influenced by the parameter λ_i. This can explain the higher cost J_{eval2} for $\lambda_1 = 3.0$ compared with the cases $\lambda_1 = 0.5$ and $\lambda_1 = 1.0$ for $\epsilon_1 \geq 0.006$, as for $\lambda_1 = 3.0$ the plant P_2 has a comparatively smaller priority according to the scheduling rule (4.15). It is worth to note that the control performance also depends on the specific reference and disturbance signals, which can explain the oscillating behavior of the cost $J_{\text{eval}i}$ with respect to ϵ_1 in Figure 4.2.

4.7 Experimental Study

Consider a set of three DC motors with the model

$$\dot{n}(t) = -\frac{k_{\text{m}}k_{\text{e}}}{2\pi JR}n(t) + \frac{r_{\text{g}}k_{\text{m}}}{2\pi JR}u(t-\tau) \tag{4.56}$$

where $n(t)$ is the speed in rounds per second, and $u(t)$ is the armature voltage in volt. There are two DC motors Maxon RE-max29 and one DC motor RS 225-9557. Table 4.4 gives the parameters. The initial motor speed of all DC motors is $n(0) = 0\,\text{s}^{-1}$. This leads to the model (4.1) with $\boldsymbol{A}_{\text{p}i} = -23.2303$, $\boldsymbol{B}_{\text{p}i} = 5.1162$ and $\boldsymbol{C}_{\text{p}i} = 1$ for $i \in \{1, 2\}$, for the DC motor Maxon RE-max29, and $\boldsymbol{A}_{\text{p}3} = -18.9440$, $\boldsymbol{B}_{\text{p}3} = 67.8754$, $\boldsymbol{C}_{\text{p}3} = 1$ for the DC motor RS 225-9557.

In this experiment the proposed online scheduler and the offline scheduler based on the sequence $\mathcal{S} := \big(3, 2, 1\big)$ are applied. As computation unit a microcontroller NXP LPC2294 is applied. Table 4.5 gives the measured execution times of the different steps for controlling the set of three DC motors under online and offline scheduling. For the input reading the online scheduling requires more time, as the output of all plants needs to be sampled in order to execute the scheduler. As under offline scheduling the scheduling sequence is pre-determined only the output of one plant needs to be sampled and the scheduling overhead is negligible. In the given setup the motor angle is measured by an encoder and transmitted via serial peripheral interface (SPI), which takes a big part of the total execution time. However, in case of the usage of different sensors this execution time can be strongly reduced, e.g. if A/D converters are used. The output

Motor	Maxon RE-max29	RS 225-9557
Motor torque constant k_{m}	$0.0258\,\frac{\text{Nm}}{\text{A}}$	$0.0444\,\frac{\text{Nm}}{\text{A}}$
Motor velocity constant k_{e}	$0.1622\,\text{Vs}$	$0.2791\,\text{Vs}$
Resistance R	$3.26\,\Omega$	$2.0611\,\Omega$
Moment of inertia J	$8.79\cdot10^{-6}\,\text{kgm}^2$	$1.5869\cdot10^{-4}\,\text{kgm}^2$
Gear ratio r_{g}	$\frac{1}{28}$	1

Table 4.4: Parameters of the DC motors

Method	Online scheduling	Offline scheduling
Input reading	626.7	208.9
Scheduling	103.6	0
Control computation	47.2	47.2
Output application	129.9	129.9
Total	907.4	386

Table 4.5: Measured worst-case execution times of the segments of the control task T_i for three DC motors in microseconds

application is realized via pulse-width modulation (PWM). The overall delay consists then of the total execution time of the segments, i.e. $\tau_i = 0.8987\,\mathrm{ms}$ and $\tau_i^{\mathrm{off}} = 0.3860\,\mathrm{ms}$ for all $i \in \mathbb{J}$. To realize a similar resource utilization the step size is selected as $h = 10\,\mathrm{ms}$ and $h^{\mathrm{off}} = \frac{10}{3}\,\mathrm{ms}$.

For determining the control parameters of both approaches the weighting matrices are selected as

$$\begin{aligned} \boldsymbol{Q}_{\mathrm{c}1/2} &= 20, & \boldsymbol{Q}_{\mathrm{I}1/2} &= 300, & \boldsymbol{R}_{\mathrm{c}1/2} &= 2, \\ \boldsymbol{Q}_{\mathrm{c}3} &= 0.2, & \boldsymbol{Q}_{\mathrm{I}3} &= 3, & \boldsymbol{R}_{\mathrm{c}3} &= 10. \end{aligned}$$

Applying Theorem 4.2 leads to the control parameters of the PI controllers for the online scheduling approach

$$\boldsymbol{K}_{1/2} = \begin{pmatrix} -0.1018 & -0.5405 \end{pmatrix}, \qquad \boldsymbol{K}_3 = \begin{pmatrix} -1.2013 & -11.7928 \end{pmatrix}.$$

Under offline scheduling the control parameters are determined applying Theorem A.1, which leads to

$$\boldsymbol{K}_{1/2}^{\mathrm{off}} = \begin{pmatrix} -0.1074 & -0.5411 \end{pmatrix}, \qquad \boldsymbol{K}_3^{\mathrm{off}} = \begin{pmatrix} -1.3265 & -11.8188 \end{pmatrix}.$$

The remaining control and scheduling parameters for the online scheduler are selected as $\lambda_i = 1$, $\epsilon_i = 0.02$ for $i \in \{1,2\}$, $\lambda_3 = 10$, $\epsilon_3 = 0.4$ and $N_i = 4$ for all $i \in \mathbb{J}$. Applying Theorem 4.3 a feasible solution is found for each plant P_i proving practical stability.

For the implementation the variables $\hat{\boldsymbol{y}}_i(0)$ and $\hat{\boldsymbol{r}}_i(0)$ are initialized as $\hat{\boldsymbol{y}}_i(0) = 0$ and $\hat{\boldsymbol{r}}_i(0) = 0$ for all $i \in \mathbb{J}_i$. Figure 4.3 shows that the output performances of both scheduling approaches are very similar. The scheduling index also shown in Figure 4.3 demonstrates how the proposed online scheduling law can distribute the resources according to the requirements. In Table 4.6 the cost J_{eval} is given for each plant P_i under both applied scheduling algorithms. For plant P_1 there is an improvement of 13.0 % under online scheduling compared with offline scheduling, for plant P_2 there is an improvement of 16.4 % and for plant P_3 a degradation of 1.6 %. This leads to an overall improvement of 8.1 % under online scheduling compared with offline scheduling. Further, the online

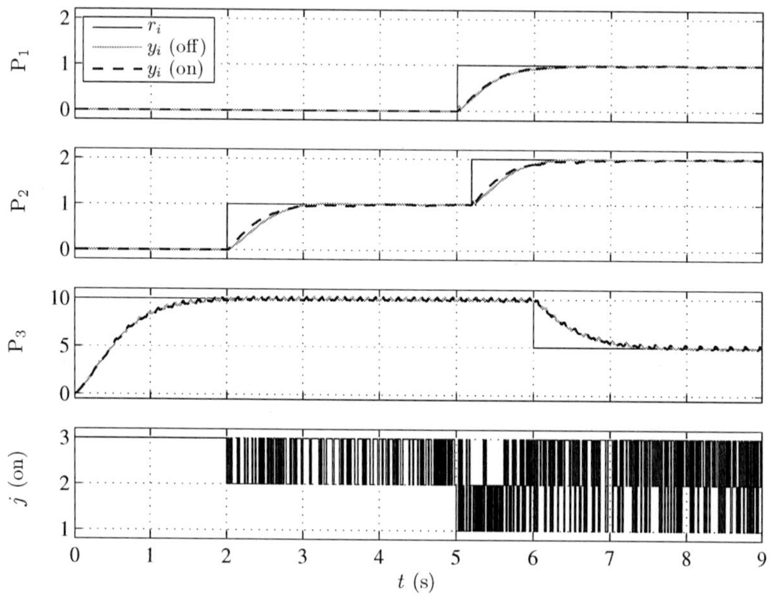

Figure 4.3: Experimental results

scheduler leads to the resource utilization is $U^{\text{on}} = 8.987\,\%$, which is smaller than under the offline scheduler ($U^{\text{off}} = 11.58\,\%$). Thus, this experimental study can emphasize the strength of the online scheduling approach distributing the resources efficiently.

Method	J_{eval1}	J_{eval2}	J_{eval3}	J_{eval}
offline scheduling	9.67	17.82	20.20	47.69
online scheduling	8.41	14.89	20.52	43.82

Table 4.6: Cost under different scheduling strategies

4.8 Summary

In this chapter a novel output-based online scheduling algorithm for PI control tasks with the aim of setpoint tracking is proposed. The scheduler distributes the limited resources online according to the requirements. Thereby, the update of the control input of plants in the steady state can be spared, such that the available resources can be used for plants in the transient phase. As this scheduling approach leads to time-varying control update intervals, the standard discrete-time PI controller is extended. For investigating the stability the closed-loop behavior of each plant is modeled as a discrete-time switched linear system. Based on a multiple Lyapunov function approach practical stability can be proved a posteriori by solving an LMI feasibility problem. The effectiveness is illustrated by simulation and practical implementation comparing the online scheduling approach with other scheduling approaches.

Part II

Event-Triggered Control

5 Event-Triggered PI Control

In this chapter, an event-triggered control concept is developed focusing on PI control. One of the main objectives is the proposition of a control synthesis, which allows to design control parameters suiting to the event-triggered implementation of the PI controller. The introduced concept equivalently allows to handle several challenges in the context of PI control such as oscillation and sticking. Further, the delay of the control input is incorporated in this work.

5.1 Problem Formulation

Consider a linear time-invariant plant controlled by a PI controller in an event-triggered manner. The plant is described by the continuous-time linear state equation

$$\begin{aligned} \dot{\boldsymbol{x}}_{\mathrm{p}}(t) &= \boldsymbol{A}_{\mathrm{p}}\boldsymbol{x}_{\mathrm{p}}(t) + \boldsymbol{B}_{\mathrm{p}}\boldsymbol{u}(t-\tau) \\ \boldsymbol{y}(t) &= \boldsymbol{C}_{\mathrm{p}}\boldsymbol{x}_{\mathrm{p}}(t) \end{aligned} \tag{5.1}$$

where $\boldsymbol{A}_{\mathrm{p}} \in \mathbb{R}^{n\times n}$ is the system matrix, $\boldsymbol{B}_{\mathrm{p}} \in \mathbb{R}^{n\times m}$ is the input matrix, $\boldsymbol{C}_{\mathrm{p}} \in \mathbb{R}^{p\times n}$ is the output matrix, $\boldsymbol{x}_{\mathrm{p}}(t) \in \mathbb{R}^{n}$ is the state vector, $\boldsymbol{y}(t) \in \mathbb{R}^{p}$ is the output vector and $\boldsymbol{u}(t-\tau) \in \mathbb{R}^{m}$ is the delayed control input vector. The input delay τ indicates the constant time delay from the measurement of the output to the actuation.

In a conventional digital control setting, the implementation of the PI controller is given by

$$\boldsymbol{x}_{\mathrm{I}}(k+1) = \boldsymbol{x}_{\mathrm{I}}(k) + h\big(\boldsymbol{y}(k) - \boldsymbol{r}\big) \tag{5.2a}$$

$$\boldsymbol{u}(k) = \boldsymbol{K}_{\mathrm{I}}\boldsymbol{x}_{\mathrm{I}}(k) + \boldsymbol{K}_{\mathrm{P}}\big(\boldsymbol{y}(k) - \boldsymbol{r}\big) \tag{5.2b}$$

where $\boldsymbol{y}(k)$ is the measured output vector at the time instant t_k, $\boldsymbol{x}_{\mathrm{I}}(k) \in \mathbb{R}^{p}$ is the integrator state, $\boldsymbol{K}_{\mathrm{P}} \in \mathbb{R}^{m\times p}$ is the proportional gain, $\boldsymbol{K}_{\mathrm{I}} \in \mathbb{R}^{m\times p}$ is the integral gain, and $\boldsymbol{r} \in \mathbb{R}^{p}$ is the constant reference signal (setpoint). The sampling period $h = t_{k+1} - t_k$ is constant.

In the following, the control of the plant is realized in a periodic event-triggered way [HDT13], which is illustrated in Figure 5.1. This means that at each time instant t_k the output vector $\boldsymbol{y}(k)$ is sampled and forwarded to the event generator (EG), where an event-triggering condition is verified. Once an event is triggered, the currently computed

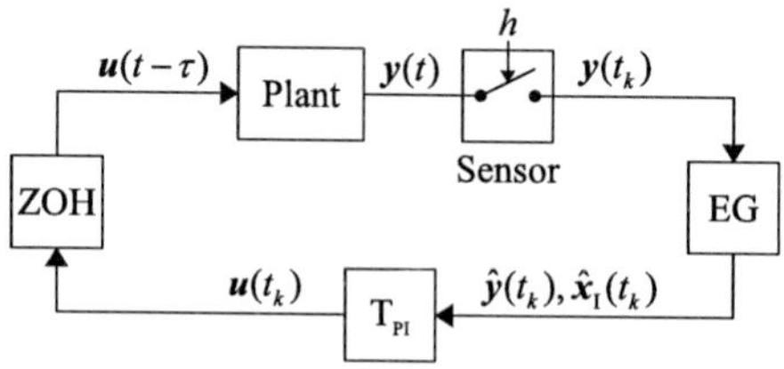

Figure 5.1: Event-triggered control architecture

integral state $\boldsymbol{x}_{\mathrm{I}}(k)$ and the measured output $\boldsymbol{y}(k)$ are forwarded to the PI control task $\mathrm{T_{PI}}$ to compute the control signal $\boldsymbol{u}(k)$ and transmit it to the plant. The control input signal is eventually updated at the time instant $t_k + \tau$ as indicated in Figure 5.2. The event generator, which will be discussed later on, requires both the measured output vector and the integrator state. Therefore, the integration of the output error $\boldsymbol{y}(k) - \boldsymbol{r}$, which is defined in (5.2a), needs to be realized in the event generator. Under the event-

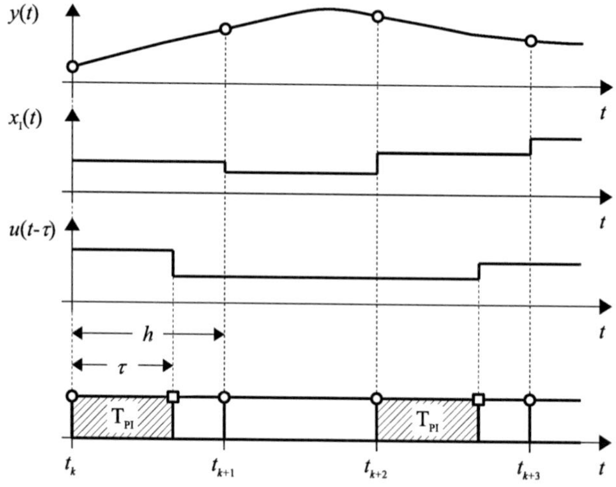

Figure 5.2: Event-triggered control timing diagram where a new measurement is indicated by a circle, a control update by a square, and the execution of a control task $\mathrm{T_{PI}}$ by the hatching

triggered control structure the PI controller is then given by

$$\boldsymbol{x}_{\mathrm{I}}(k+1) = \boldsymbol{x}_{\mathrm{I}}(t_k) + h\big(\boldsymbol{y}(k) - \boldsymbol{r}\big) \tag{5.3a}$$

$$\boldsymbol{u}(k) = \boldsymbol{K}_{\mathrm{I}}\hat{\boldsymbol{x}}_{\mathrm{I}}(k) + \boldsymbol{K}_{\mathrm{P}}\big(\hat{\boldsymbol{y}}(k) - \boldsymbol{r}\big) \tag{5.3b}$$

where the variables $\hat{\boldsymbol{y}}(t_k)$ and $\hat{\boldsymbol{x}}_{\mathrm{I}}(t_k)$ are given by

$$\begin{pmatrix} \hat{\boldsymbol{y}}(k) \\ \hat{\boldsymbol{x}}_{\mathrm{I}}(k) \end{pmatrix} = \begin{cases} \begin{pmatrix} \boldsymbol{y}(k) \\ \boldsymbol{x}_{\mathrm{I}}(k) \end{pmatrix} & \text{if an event is triggered} \\ \begin{pmatrix} \hat{\boldsymbol{y}}(k-1) \\ \hat{\boldsymbol{x}}_{\mathrm{I}}(k-1) \end{pmatrix} & \text{otherwise.} \end{cases} \tag{5.4}$$

Thus, the values of $\hat{\boldsymbol{y}}(k)$ and $\hat{\boldsymbol{x}}_{\mathrm{I}}(k)$ in Figure 5.1 indicate the most recently transmitted values of the output vector and the integrator state.

Because of the digital implementation platform, a discrete-time model of (5.1) is derived and considered for further analysis. The state equation is discretized over the sampling interal $t_k \leq t < t_{k+1}$ using zero-order hold (ZOH). The delayed control vector is given as

$$\boldsymbol{u}(t-\tau) = \begin{cases} \boldsymbol{u}(t_{k-1}) & \text{for} \quad t_k \leq t < t_k + \tau \\ \boldsymbol{u}(t_k) & \text{for} \quad t_k + \tau \leq t < t_{k+1}. \end{cases} \tag{5.5}$$

Considering the delayed control signal, an augmented discrete-time state equation corresponding to (5.1) is formulated as

$$\begin{aligned} \boldsymbol{x}_{\mathrm{a}}(k+1) &= \boldsymbol{A}_{\mathrm{a}}\boldsymbol{x}_{\mathrm{a}}(k) + \boldsymbol{B}_{\mathrm{a}}\boldsymbol{u}(k) \\ \boldsymbol{y}(k) &= \boldsymbol{C}_{\mathrm{a}}\boldsymbol{x}_{\mathrm{a}}(k) \end{aligned} \tag{5.6}$$

with $\boldsymbol{x}_{\mathrm{a}}(k) = \begin{pmatrix} \boldsymbol{x}_{\mathrm{p}}(k)^T & \boldsymbol{u}(k-1) \end{pmatrix}^T$ and

$$\boldsymbol{A}_{\mathrm{a}} = \begin{pmatrix} e^{\boldsymbol{A}_{\mathrm{p}}h} & \int_{h-\tau}^{h} e^{\boldsymbol{A}_{\mathrm{p}}s} ds \boldsymbol{B}_{\mathrm{p}} \\ \boldsymbol{0} & \boldsymbol{0} \end{pmatrix}, \quad \boldsymbol{B}_{\mathrm{a}} = \begin{pmatrix} \int_{0}^{h-\tau} e^{\boldsymbol{A}_{\mathrm{p}}s} ds \boldsymbol{B}_{\mathrm{p}} \\ \boldsymbol{I} \end{pmatrix}, \quad \boldsymbol{C}_{\mathrm{a}} = \begin{pmatrix} \boldsymbol{C}_{\mathrm{p}} & \boldsymbol{0} \end{pmatrix}.$$

For a detailed derivation of the discrete-time state space matrices of (5.6) see [ÅW90, Section 3.2].

In the following, the goal is to propose a control synthesis which allows designing the PI control parameters taking the event-triggered implementation into account. In order to quantify the control performance the continuous-time quadratic cost function

$$J = \int_0^\infty \begin{pmatrix} \tilde{\boldsymbol{x}}_{\mathrm{p}}(t) \\ \tilde{\boldsymbol{x}}_{\mathrm{I}}(t) \\ \tilde{\boldsymbol{u}}(t-\tau) \end{pmatrix}^T \begin{pmatrix} \boldsymbol{Q}_{\mathrm{c}} & 0 & 0 \\ 0 & \boldsymbol{Q}_{\mathrm{I}} & 0 \\ 0 & 0 & \boldsymbol{R}_{\mathrm{c}} \end{pmatrix} \begin{pmatrix} \tilde{\boldsymbol{x}}_{\mathrm{p}}(t) \\ \tilde{\boldsymbol{x}}_{\mathrm{I}}(t) \\ \tilde{\boldsymbol{u}}(t-\tau) \end{pmatrix} dt \tag{5.7}$$

with $\boldsymbol{Q}_{\mathrm{c}} \in \mathbb{R}^{n\times n}$ and $\boldsymbol{Q}_{\mathrm{I}} \in \mathbb{R}^{p\times p}$ symmetric and positive semidefinite and $\boldsymbol{R}_{\mathrm{c}} \in \mathbb{R}^{m\times m}$ symmetric and positive definite is introduced. The deviation variables

$$\tilde{\boldsymbol{x}}_{\mathrm{p}}(t) = \boldsymbol{x}_{\mathrm{p}}(t) - \boldsymbol{x}_{\mathrm{p}}^{\mathrm{ss}} \tag{5.8a}$$

$$\tilde{\boldsymbol{u}}(t-\tau) = \boldsymbol{u}(t-\tau) - \boldsymbol{u}^{\mathrm{ss}} \tag{5.8b}$$

$$\tilde{\boldsymbol{x}}_{\mathrm{I}}(t) = \boldsymbol{x}_{\mathrm{I}}(t) - \boldsymbol{x}_{\mathrm{I}}^{\mathrm{ss}} \tag{5.8c}$$

are defined based on the steady-state values $\boldsymbol{x}_{\mathrm{P}}^{\mathrm{ss}}$, $\boldsymbol{u}^{\mathrm{ss}}$ and $\boldsymbol{x}_{\mathrm{I}}^{\mathrm{ss}}$ that are determined by

$$\begin{pmatrix} \boldsymbol{x}_{\mathrm{P}}^{\mathrm{ss}} \\ \boldsymbol{x}_{\mathrm{I}}^{\mathrm{ss}} \end{pmatrix} = \begin{pmatrix} \boldsymbol{A}_{\mathrm{P}} & \boldsymbol{B}_{\mathrm{P}}\boldsymbol{K}_{\mathrm{I}} \\ \boldsymbol{C}_{\mathrm{P}} & \boldsymbol{0} \end{pmatrix}^{-1} \begin{pmatrix} \boldsymbol{0} \\ \boldsymbol{r} \end{pmatrix} \tag{5.9a}$$

$$\boldsymbol{u}^{\mathrm{ss}} = \boldsymbol{K}_{\mathrm{I}}\boldsymbol{x}_{\mathrm{I}}^{\mathrm{ss}} \tag{5.9b}$$

which can be derived from (5.1) and (5.3b) setting $\dot{\boldsymbol{x}}_{\mathrm{P}}(t) = \boldsymbol{0}$ and $\boldsymbol{y}(t) = \boldsymbol{r}$.

The continuous-time cost function (5.7) is discretized analogously to the continuous-time state equation (5.1). Thereby, it needs to be taken into account that the integrator state $\boldsymbol{x}_{\mathrm{I}}(t)$ is a piecewise constant state, i.e. $\boldsymbol{x}_{\mathrm{I}}(t) = \boldsymbol{x}_{\mathrm{I}}(k)$ for $t \in [t_k, t_{k+1})$. The discretized cost function associated to the system (5.1) can then be rewritten as

$$J = \sum_{k=0}^{\infty} \begin{pmatrix} \tilde{\boldsymbol{x}}_{\mathrm{P}}(k) \\ \tilde{\boldsymbol{u}}(k-1) \\ \tilde{\boldsymbol{x}}_{\mathrm{I}}(k) \\ \tilde{\boldsymbol{u}}(k) \end{pmatrix}^T \boldsymbol{Q} \begin{pmatrix} \tilde{\boldsymbol{x}}_{\mathrm{P}}(k) \\ \tilde{\boldsymbol{u}}(k-1) \\ \tilde{\boldsymbol{x}}_{\mathrm{I}}(k) \\ \tilde{\boldsymbol{u}}(k) \end{pmatrix} \tag{5.10}$$

with deviation variables defined analog to (5.8). The discretization procedure for calculating the weighting matrix $\boldsymbol{Q}$ is explained in the Appendix A.1.3.

5.2 Event-Triggered PI Control Synthesis

In this section the control synthesis is presented with respect to two different event-triggering conditions. One event-triggering condition is based on the control input error [DH12, HDT13, WRGL15] given by

$$\left\| \boldsymbol{K} \begin{pmatrix} \hat{\boldsymbol{y}}(k-1) \\ \hat{\boldsymbol{x}}_{\mathrm{I}}(k-1) \end{pmatrix} - \boldsymbol{K} \begin{pmatrix} \boldsymbol{y}(k) \\ \boldsymbol{x}_{\mathrm{I}}(k) \end{pmatrix} \right\|_2^2 > \sigma^2 \left\| \boldsymbol{K} \begin{pmatrix} \boldsymbol{y}(k) - \boldsymbol{r} \\ \boldsymbol{x}_{\mathrm{I}}(k) \end{pmatrix} - \boldsymbol{u}^{\mathrm{ss}} \right\|_2^2 + \epsilon \tag{5.11}$$

with $\boldsymbol{K} = \begin{pmatrix} \boldsymbol{K}_{\mathrm{P}} & \boldsymbol{K}_{\mathrm{I}} \end{pmatrix}$ and the scalars $\sigma \in \mathbb{R}_0^+$ and $\epsilon \in \mathbb{R}_0^+$ as design parameters. The left-hand side of the event-triggering condition (5.11) accounts the deviation of the control input based on the actual measurements from the control input based on the most recently transmitted variables $\hat{\boldsymbol{x}}_{\mathrm{I}}(t_k)$ and $\hat{\boldsymbol{y}}(t_k)$. This event-triggering condition implies that if the deviation of the control input has only a minor change, measured by the threshold on the right-hand side, it can still be very effective.

Another event-triggering condition based on the output is given by

$$\left\| \begin{pmatrix} \hat{\boldsymbol{y}}(k-1) - \boldsymbol{y}(k) \\ \hat{\boldsymbol{x}}_{\mathrm{I}}(k-1) - \boldsymbol{x}_{\mathrm{I}}(k) \end{pmatrix} \right\|_2^2 > \sigma^2 \left\| \begin{pmatrix} \boldsymbol{y}(k) - \boldsymbol{r} \\ \boldsymbol{x}_{\mathrm{I}}(k) - \boldsymbol{x}_{\mathrm{I}}^{\mathrm{ss}} \end{pmatrix} \right\|_2^2 + \epsilon \tag{5.12}$$

with $\sigma \in \mathbb{R}_0^+$ and $\epsilon \in \mathbb{R}_0^+$. This event-triggering condition is motivated by the fact that under a small variation of the output and integrator state the previous control input can still be effective.

For the control synthesis, the parameter ϵ is set zero. With $\epsilon = 0$ global asymptotic stability can be guaranteed by selecting σ suitably, which is crucial for the control synthesis. On the other hand, the adding of a small value for ϵ allows to reduce significantly the number of triggered events with only a marginal decrease of the control performance, see the evaluation in Section 5.4 and 5.5. However, for $\epsilon \neq 0$ asymptotic stability cannot be achieved but only practical stability, which is analyzed after the control synthesis in Section 5.3. The PI control synthesis problem then is formulated as follows.

Problem 5.1 *For the discrete-time system* (5.6) *and a given tuning parameter* $\sigma \in \mathbb{R}_0^+$ *find the control gain* $\boldsymbol{K} = \begin{pmatrix}\boldsymbol{K}_\mathrm{P} & \boldsymbol{K}_\mathrm{I}\end{pmatrix}$ *such that the closed-loop system under the control law* (5.3), (5.4) *and the event-triggering condition* (5.11) *and* (5.12), *respectively, with* $\epsilon = 0$ *is globally asymptotically stable (GAS) with guaranteed performance.*

Without loss of generality, in the following derivations it is assumed that the reference signal is $\boldsymbol{r} = \boldsymbol{0}$. Thus, all steady state values $\boldsymbol{x}_\mathrm{p}^\mathrm{ss}$, $\boldsymbol{x}_\mathrm{I}^\mathrm{ss}$ and $\boldsymbol{u}^\mathrm{ss}$ are zero.

5.2.1 Control Synthesis for Control Input Based Event-Triggered PI Control

Considering the event triggering condition (5.11) the closed-loop system is built in order to derive a stability condition, which is then extended to the control synthesis. Introducing the error variable

$$\boldsymbol{e}_\mathrm{u}(k) = \boldsymbol{K}\begin{pmatrix}\hat{\boldsymbol{y}}(k)\\ \hat{\boldsymbol{x}}_\mathrm{I}(k)\end{pmatrix} - \boldsymbol{K}\begin{pmatrix}\boldsymbol{y}(k)\\ \boldsymbol{x}_\mathrm{I}(k)\end{pmatrix} \tag{5.13}$$

and substituting the discrete-time PI controller (5.3) into the discrete-time state equation (5.6) yields

$$\boldsymbol{z}(k+1) = (\boldsymbol{A}_\mathrm{z} + \boldsymbol{B}_\mathrm{z}\boldsymbol{K}\boldsymbol{C}_\mathrm{z})\,\boldsymbol{z}(k) + \boldsymbol{B}_\mathrm{z}\boldsymbol{e}_\mathrm{u}(k) \tag{5.14}$$

with $\boldsymbol{z}(k) = \begin{pmatrix}\boldsymbol{x}_\mathrm{a}^T(k) & \boldsymbol{x}_\mathrm{I}^T(k)\end{pmatrix}^T$ and

$$\boldsymbol{A}_\mathrm{z} = \begin{pmatrix}\boldsymbol{A}_\mathrm{a} & 0\\ h\boldsymbol{C}_\mathrm{a} & \boldsymbol{I}\end{pmatrix}, \quad \boldsymbol{B}_\mathrm{z} = \begin{pmatrix}\boldsymbol{B}_\mathrm{a}\\ 0\end{pmatrix}, \quad \boldsymbol{C}_\mathrm{z} = \begin{pmatrix}\boldsymbol{C}_\mathrm{a} & 0\\ 0 & \boldsymbol{I}\end{pmatrix}. \tag{5.15}$$

Similarly, the discrete-time PI controller (5.3) and the error variable (5.13) are substituted into the discrete-time cost function (5.10) resulting in the discrete-time closed loop cost function

$$J = \sum_{k=0}^{\infty}\begin{pmatrix}\boldsymbol{z}(k)\\ \boldsymbol{e}_\mathrm{u}(k)\end{pmatrix}^T \tilde{\boldsymbol{Q}}\begin{pmatrix}\boldsymbol{z}(k)\\ \boldsymbol{e}_\mathrm{u}(k)\end{pmatrix} \tag{5.16}$$

where

$$\tilde{\boldsymbol{Q}} = \begin{pmatrix}\boldsymbol{I} & (\boldsymbol{K}\boldsymbol{C}_\mathrm{z})^T\\ 0 & \boldsymbol{I}\end{pmatrix}\boldsymbol{Q}\begin{pmatrix}\boldsymbol{I} & 0\\ \boldsymbol{K}\boldsymbol{C}_\mathrm{z} & \boldsymbol{I}\end{pmatrix}. \tag{5.17}$$

It is worth to note that the resulting weighting matrix $\tilde{\boldsymbol{Q}}$ is symmetric and positive semidefinite, a property used in the following.

For analyzing the stability, the quadratic Lyapunov function

$$V(k) = \boldsymbol{z}^T(k)\boldsymbol{P}\boldsymbol{z}(k) \tag{5.18}$$

with $\boldsymbol{P} \in \mathbb{R}^{(n+m+p)\times(n+m+p)}$ symmetric and positive definite is introduced. The difference of the Lyapunov function $\Delta V(k) = V(k+1) - V(k)$ along the trajectories of the closed-loop system (5.14) is given by

$$\Delta V(k) = \begin{pmatrix} \boldsymbol{z}(k) \\ \boldsymbol{e}_\mathrm{u}(k) \end{pmatrix}^T \begin{pmatrix} \tilde{\boldsymbol{A}}_\mathrm{z}^T \boldsymbol{P} \tilde{\boldsymbol{A}}_\mathrm{z} - \boldsymbol{P} & * \\ \boldsymbol{B}_\mathrm{z}^T \boldsymbol{P} \tilde{\boldsymbol{A}}_\mathrm{z} & \boldsymbol{B}_\mathrm{z}^T \boldsymbol{P} \boldsymbol{B}_\mathrm{z} \end{pmatrix} \begin{pmatrix} \boldsymbol{z}(k) \\ \boldsymbol{e}_\mathrm{u}(k) \end{pmatrix} \tag{5.19}$$

where $\tilde{\boldsymbol{A}}_\mathrm{z} = \boldsymbol{A}_\mathrm{z} + \boldsymbol{B}_\mathrm{z}\boldsymbol{K}\boldsymbol{C}_\mathrm{z}$. First, a theorem is presented based on which stability can be analyzed for given control and event parameters $\boldsymbol{K}$ and σ.

Theorem 5.1 *The discrete-time closed-loop system* (5.14) *is GAS under event-triggered PI control* (5.3), (5.4), (5.11) *if the LMI feasibility problem*

$$\begin{pmatrix} \boldsymbol{P} - \tilde{\boldsymbol{A}}_\mathrm{z}^T \boldsymbol{P} \tilde{\boldsymbol{A}}_\mathrm{z} & * \\ -\boldsymbol{B}_\mathrm{z}^T \boldsymbol{P} \tilde{\boldsymbol{A}}_\mathrm{z} & -\boldsymbol{B}_\mathrm{z}^T \boldsymbol{P} \boldsymbol{B}_\mathrm{z} \end{pmatrix} - \mu \begin{pmatrix} \sigma^2 \boldsymbol{C}_\mathrm{z}^T \boldsymbol{K}^T \boldsymbol{K} \boldsymbol{C}_\mathrm{z} & \boldsymbol{0} \\ \boldsymbol{0} & -\boldsymbol{I} \end{pmatrix} - \tilde{\boldsymbol{Q}} > 0 \tag{5.20}$$

holds with the LMI variables $\boldsymbol{P} \in \mathbb{R}^{(n+m+p)\times(n+m+p)}$ symmetric and positive definite and the scalar $\mu \in \mathbb{R}_0^+$. Furthermore, the closed-loop cost function (5.16) *is bounded.*

PROOF. Assume that there exists a matrix $\boldsymbol{P} > \boldsymbol{0}$ such that the LMI condition (5.20) is feasible. Pre-/post-multiplying (5.20) by the vector $\begin{pmatrix} \boldsymbol{z}^T(k) & \boldsymbol{e}_\mathrm{u}^T(k) \end{pmatrix}$ and its transpose, respectively, yields

$$\Delta V(k) < -\mu \begin{pmatrix} \boldsymbol{z}(k) \\ \boldsymbol{e}_\mathrm{u}(k) \end{pmatrix}^T \begin{pmatrix} \sigma^2 \boldsymbol{C}_\mathrm{z}^T \boldsymbol{K}^T \boldsymbol{K} \boldsymbol{C}_\mathrm{z} & \boldsymbol{0} \\ \boldsymbol{0} & -\boldsymbol{I} \end{pmatrix} \begin{pmatrix} \boldsymbol{z}(k) \\ \boldsymbol{e}_\mathrm{u}(k) \end{pmatrix} - \begin{pmatrix} \boldsymbol{z}(k) \\ \boldsymbol{e}_\mathrm{u}(k) \end{pmatrix}^T \tilde{\boldsymbol{Q}} \begin{pmatrix} \boldsymbol{z}(k) \\ \boldsymbol{e}_\mathrm{u}(k) \end{pmatrix}. \tag{5.21}$$

Considering the definition of $\hat{\boldsymbol{y}}(k)$ and $\hat{\boldsymbol{x}}_\mathrm{I}(k)$ (5.4) and the event-triggering condition (5.11) a bound on the error variable $\boldsymbol{e}_\mathrm{u}(k)$ defined by (5.13) can be given as

$$\boldsymbol{e}_\mathrm{u}^T(k)\boldsymbol{e}_\mathrm{u}(k) \leq \sigma^2 \boldsymbol{z}^T(k) \boldsymbol{C}_\mathrm{z}^T \boldsymbol{K}^T \boldsymbol{K} \boldsymbol{C}_\mathrm{z} \boldsymbol{z}(k). \tag{5.22}$$

Based on the bound (5.22) and the constraints $\mu \geq 0$ and $\tilde{\boldsymbol{Q}} \geq \boldsymbol{0}$, the right-hand side of inequality (5.21) is not larger than zero, guaranteeing the negative definiteness of $\Delta V(k)$, which implies that (5.14) is GAS. Due to the bound (5.22) it can be concluded from (5.21) that

$$\Delta V(k) < -\begin{pmatrix} \boldsymbol{z}(k) \\ \boldsymbol{e}_\mathrm{u}(k) \end{pmatrix}^T \tilde{\boldsymbol{Q}} \begin{pmatrix} \boldsymbol{z}(k) \\ \boldsymbol{e}_\mathrm{u}(k) \end{pmatrix} \tag{5.23}$$

always holds. Summing up (5.23) over $k = 0, ..., \infty$ yields

$$V(0) - \lim_{k\to\infty} V(k) > J. \tag{5.24}$$

As the closed-loop system (5.14) is GAS, $V(k) \to 0$ for $k \to \infty$. Thus, inequality (5.24) implies an upper bound

$$J < \boldsymbol{z}^T(0)\boldsymbol{P}\boldsymbol{z}(0) < \mathrm{tr}(\boldsymbol{P})\,\|\boldsymbol{z}(0)\|_2^2\,. \tag{5.25}$$

This completes the proof. □

In order to minimize the upper bound given in (5.25), Theorem 5.1 is extended to the following corollary.

Corollary 5.1 *Under event-triggered PI control* (5.3), (5.4), (5.11) *with given parameters* $\boldsymbol{K}$ *and* σ *a minimum upper bound of the cost function* (5.16) *is obtained from the LMI optimization problem*

$$\min_{\boldsymbol{P}} \mathrm{tr}\,(\boldsymbol{P}) \quad \text{subject to} \tag{5.26a}$$

$$\begin{pmatrix} \boldsymbol{P} - \tilde{\boldsymbol{A}}_{\mathrm{z}}^T\boldsymbol{P}\tilde{\boldsymbol{A}}_{\mathrm{z}} & * \\ -\boldsymbol{B}_{\mathrm{z}}^T\boldsymbol{P}\tilde{\boldsymbol{A}}_{\mathrm{z}} & -\boldsymbol{B}_{\mathrm{z}}^T\boldsymbol{P}\boldsymbol{B}_{\mathrm{z}} \end{pmatrix} - \mu \begin{pmatrix} \sigma^2\boldsymbol{C}_{\mathrm{z}}^T\boldsymbol{K}^T\boldsymbol{K}\boldsymbol{C}_{\mathrm{z}} & 0 \\ 0 & -\boldsymbol{I} \end{pmatrix} - \tilde{\boldsymbol{Q}} > 0 \tag{5.26b}$$

with the LMI variables $\boldsymbol{P} \in \mathbb{R}^{(n+m+p)\times(n+m+p)}$ *symmetric and positive definite and the scalar* $\mu \in \mathbb{R}_0^+$.

Theorem 5.1 is now extended to solve the control synthesis problem.

Theorem 5.2 *A solution to Problem 5.1 with respect to the event-triggering condition* (5.11) *is obtained from the LMI optimization problem*

$$\min_{\boldsymbol{Z},\boldsymbol{K}} \mathrm{tr}\left(\boldsymbol{Z}^{-1}\right) \quad \text{subject to} \tag{5.27a}$$

$$\boldsymbol{V}\boldsymbol{C}_{\mathrm{z}} = \boldsymbol{C}_{\mathrm{z}}\boldsymbol{G} \tag{5.27b}$$

$$\begin{pmatrix} \boldsymbol{G}^T + \boldsymbol{G} - \boldsymbol{Z} & * & * & * & * \\ 0 & \kappa\boldsymbol{I} & * & * & * \\ \boldsymbol{A}_{\mathrm{z}}\boldsymbol{G} + \boldsymbol{B}_{\mathrm{z}}\boldsymbol{U}\boldsymbol{C}_{\mathrm{z}} & \kappa\boldsymbol{B}_{\mathrm{z}} & \boldsymbol{Z} & * & * \\ \sigma\boldsymbol{U}\boldsymbol{C}_{\mathrm{z}} & 0 & 0 & \kappa\boldsymbol{I} & * \\ \boldsymbol{Q}^{1/2}\begin{pmatrix} \boldsymbol{G} \\ \boldsymbol{U}\boldsymbol{C}_{\mathrm{z}} \end{pmatrix} & \kappa\boldsymbol{Q}^{1/2}\begin{pmatrix} 0 \\ \boldsymbol{I} \end{pmatrix} & 0 & 0 & \boldsymbol{I} \end{pmatrix} > 0 \tag{5.27c}$$

with the LMI variables $\boldsymbol{G} \in \mathbb{R}^{(n+m+p)\times(n+m+p)}$, $\boldsymbol{V} \in \mathbb{R}^{2p\times 2p}$ *and* $\boldsymbol{U} \in \mathbb{R}^{m\times 2p}$ *unrestricted,* $\boldsymbol{Z} = \boldsymbol{P}^{-1}$ *symmetric positive definite, and the positive scalar* $\kappa \in \mathbb{R}^+$. *The control gain results from*

$$\boldsymbol{K} = \boldsymbol{U}\boldsymbol{V}^{-1}. \tag{5.28}$$

PROOF. Assume that the conditions (5.27a) and (5.27b) are satisfied and consider further that $\boldsymbol{C}_\mathrm{z}$ is of full-row rank. Then (5.27b) implies that $\boldsymbol{V}$ is of full rank and thus invertible [DRI02]. Combining (5.27b) and (5.28) leads to

$$\boldsymbol{U}\boldsymbol{C}_\mathrm{z} = \boldsymbol{K}\boldsymbol{V}\boldsymbol{C}_\mathrm{z} = \boldsymbol{K}\boldsymbol{C}_\mathrm{z}\boldsymbol{G}. \tag{5.29}$$

Since $\boldsymbol{Z} > \boldsymbol{0}$, also

$$(\boldsymbol{Z} - \boldsymbol{G})^T \boldsymbol{Z}^{-1} (\boldsymbol{Z} - \boldsymbol{G}) \geq 0 \tag{5.30}$$

holds. Inequality (5.30) is equivalent to

$$\boldsymbol{G}^T \boldsymbol{Z}^{-1} \boldsymbol{G} \geq \boldsymbol{G}^T + \boldsymbol{G} - \boldsymbol{Z}. \tag{5.31}$$

Based on (5.29) and (5.31) inequality (5.27c) implies

$$\begin{pmatrix} \boldsymbol{G}^T\boldsymbol{Z}^{-1}\boldsymbol{G} & * & * & * & * \\ 0 & \kappa\boldsymbol{I} & * & * & * \\ \boldsymbol{A}_\mathrm{z}\boldsymbol{G} + \boldsymbol{B}_\mathrm{z}\boldsymbol{K}\boldsymbol{C}_\mathrm{z}\boldsymbol{G} & \kappa\boldsymbol{B}_\mathrm{z} & \boldsymbol{Z} & * & * \\ \sigma\boldsymbol{K}\boldsymbol{C}_\mathrm{z}\boldsymbol{G} & 0 & 0 & \kappa\boldsymbol{I} & * \\ \boldsymbol{Q}^{1/2}\begin{pmatrix} \boldsymbol{G} \\ \boldsymbol{K}\boldsymbol{C}_\mathrm{z}\boldsymbol{G} \end{pmatrix} & \kappa\boldsymbol{Q}^{1/2}\begin{pmatrix} 0 \\ \boldsymbol{I} \end{pmatrix} & 0 & 0 & \boldsymbol{I} \end{pmatrix} > 0. \tag{5.32}$$

Pre-/post-multiplying (5.32) by $\operatorname{diag}(\boldsymbol{G}^{-T}, \kappa^{-1}\boldsymbol{I}, \boldsymbol{I}, \boldsymbol{I}, \boldsymbol{I})$ and $\operatorname{diag}(\boldsymbol{G}^{-1}, \kappa^{-1}\boldsymbol{I}, \boldsymbol{I}, \boldsymbol{I}, \boldsymbol{I})$, respectively, results in

$$\begin{pmatrix} \boldsymbol{Z}^{-1} & * & * & * & * \\ 0 & \kappa^{-1}\boldsymbol{I} & * & * & * \\ \boldsymbol{A}_\mathrm{z} + \boldsymbol{B}_\mathrm{z}\boldsymbol{K}\boldsymbol{C}_\mathrm{z} & \boldsymbol{B}_\mathrm{z} & \boldsymbol{Z} & * & * \\ \sigma\boldsymbol{K}\boldsymbol{C}_\mathrm{z} & 0 & 0 & \kappa\boldsymbol{I} & * \\ \boldsymbol{Q}^{1/2}\begin{pmatrix} \boldsymbol{I} \\ \boldsymbol{K}\boldsymbol{C}_\mathrm{z} \end{pmatrix} & \boldsymbol{Q}^{1/2}\begin{pmatrix} 0 \\ \boldsymbol{I} \end{pmatrix} & 0 & 0 & \boldsymbol{I} \end{pmatrix} > 0. \tag{5.33}$$

Substituting $\boldsymbol{P} = \boldsymbol{Z}^{-1}$ and applying the Schur complement twice leads to

$$\begin{pmatrix} \boldsymbol{P} - \tilde{\boldsymbol{A}}_\mathrm{z}^T\boldsymbol{P}\tilde{\boldsymbol{A}}_\mathrm{z} & * \\ -\boldsymbol{B}_\mathrm{z}^T\boldsymbol{P}\tilde{\boldsymbol{A}}_\mathrm{z} & -\boldsymbol{B}_\mathrm{z}^T\boldsymbol{P}\boldsymbol{B}_\mathrm{z} \end{pmatrix} - \kappa^{-1}\begin{pmatrix} \sigma^2\boldsymbol{C}_\mathrm{z}^T\boldsymbol{K}^T\boldsymbol{K}\boldsymbol{C}_\mathrm{z} & 0 \\ 0 & -\boldsymbol{I} \end{pmatrix} - \tilde{\boldsymbol{Q}} > 0 \tag{5.34}$$

Based on Theorem 5.1 it follows from (5.34) that the closed-loop system (5.14) is GAS by substituting $\mu = \kappa^{-1}$.

By minimizing the objective function (5.27a), the upper bound (5.25) is minimized, leading to a guaranteed closed-loop cost

$$J < \operatorname{tr}(\boldsymbol{P}) \left\| \boldsymbol{z}(0) \right\|_2^2. \tag{5.35}$$

This completes the proof. □

Remark 5.1. For an output-based optimal control design the optimal control gain matrix depends on the initial state, see [LVS12, Section 8.1]. As the initial state is usually unknown, the aim is to design a controller for an average initial state. Therefore, the initial state is assumed to be a Gaussian random variable with mean $\mathbf{0}$ and covariance matrix $\boldsymbol{I}$. Then the expected value of $\boldsymbol{z}^T(0)\boldsymbol{P}\boldsymbol{z}(0)$ is given by

$$\mathrm{E}\left(\boldsymbol{z}^T(0)\boldsymbol{P}\boldsymbol{z}(0)\right) = \mathrm{tr}(\boldsymbol{P}), \tag{5.36}$$

where $\mathrm{E}(\cdot)$ denotes the expected value, which motivates the objective function in Theorem 5.2, see also [ÅW90, page 338].

5.2.2 Control Synthesis for Output Based Event-Triggered PI Control

In the following, similar derivations as given in Section 5.2.1 are made for the output based event-triggering condition (5.12). First, the error variable

$$\boldsymbol{e}_{\mathrm{y}}(k) = \begin{pmatrix} \hat{\boldsymbol{y}}(k) \\ \hat{\boldsymbol{x}}_{\mathrm{I}}(k) \end{pmatrix} - \begin{pmatrix} \boldsymbol{y}(k) \\ \boldsymbol{x}_{\mathrm{I}}(k) \end{pmatrix} \tag{5.37}$$

is introduced. Substituting (5.37) and the discrete-time PI controller (5.3) into the discrete-time state equation (5.6) results in

$$\boldsymbol{z}(k+1) = \left(\boldsymbol{A}_{\mathrm{z}} + \boldsymbol{B}_{\mathrm{z}}\boldsymbol{K}\boldsymbol{C}_{\mathrm{z}}\right)\boldsymbol{z}(k) + \left(\boldsymbol{B}_{\mathrm{z}}\boldsymbol{K}\right)\boldsymbol{e}_{\mathrm{y}}(k) \tag{5.38}$$

with the matrices defined according to (5.15). Substituting the discrete-time PI controller (5.3) and the error variable (5.37) into the discrete-time cost function (5.10) results in the discrete-time closed loop cost function

$$J = \sum_{k=0}^{\infty} \begin{pmatrix} \boldsymbol{z}(k) \\ \boldsymbol{e}_{\mathrm{y}}(k) \end{pmatrix}^T \tilde{\boldsymbol{Q}} \begin{pmatrix} \boldsymbol{z}(k) \\ \boldsymbol{e}_{\mathrm{y}}(k) \end{pmatrix} \tag{5.39}$$

where

$$\tilde{\boldsymbol{Q}} = \begin{pmatrix} \boldsymbol{I} & (\boldsymbol{K}\boldsymbol{C}_{\mathrm{z}})^T \\ 0 & \boldsymbol{K}^T \end{pmatrix} \boldsymbol{Q} \begin{pmatrix} \boldsymbol{I} & 0 \\ \boldsymbol{K}\boldsymbol{C}_{\mathrm{z}} & \boldsymbol{K} \end{pmatrix}. \tag{5.40}$$

The resulting weighting matrix $\tilde{\boldsymbol{Q}}$ is symmetric and positive semidefinite, which is used in the following.

Under the same quadratic Lyapunov function (5.18) the difference of the Lyapunov function $\Delta V(k) = V(k+1) - V(k)$ along the trajectories of the closed-loop system (5.39) is given by

$$\Delta V(k) = \begin{pmatrix} \boldsymbol{z}(k) \\ \boldsymbol{e}_{\mathrm{y}}(k) \end{pmatrix}^T \begin{pmatrix} \tilde{\boldsymbol{A}}_{\mathrm{z}}^T\boldsymbol{P}\tilde{\boldsymbol{A}}_{\mathrm{z}} - \boldsymbol{P} & * \\ \tilde{\boldsymbol{B}}_{\mathrm{z}}^T\boldsymbol{P}\tilde{\boldsymbol{A}}_{\mathrm{z}} & \tilde{\boldsymbol{B}}_{\mathrm{z}}^T\boldsymbol{P}\tilde{\boldsymbol{B}}_{\mathrm{z}} \end{pmatrix} \begin{pmatrix} \boldsymbol{z}(k) \\ \boldsymbol{e}_{\mathrm{y}}(k) \end{pmatrix} \tag{5.41}$$

where $\tilde{\boldsymbol{A}}_{\mathrm{z}} = \boldsymbol{A}_{\mathrm{z}} + \boldsymbol{B}_{\mathrm{z}}\boldsymbol{K}\boldsymbol{C}_{\mathrm{z}}$ and $\tilde{\boldsymbol{B}}_{\mathrm{z}} = \boldsymbol{B}_{\mathrm{z}}\boldsymbol{K}$. For given parameters $\boldsymbol{K}$ and σ the stability can then be analyzed based on the following theorem.

Theorem 5.3 *The discrete-time closed-loop system* (5.38) *is GAS under event-triggered PI control* (5.3), (5.4), (5.12) *if the LMI feasibility problem*

$$\begin{pmatrix} \boldsymbol{P}-\tilde{\boldsymbol{A}}_{\mathrm{z}}^T\boldsymbol{P}\tilde{\boldsymbol{A}}_{\mathrm{z}} & * \\ -\tilde{\boldsymbol{B}}_{\mathrm{z}}^T\boldsymbol{P}\tilde{\boldsymbol{A}}_{\mathrm{z}} & -\tilde{\boldsymbol{B}}_{\mathrm{z}}^T\boldsymbol{P}\tilde{\boldsymbol{B}}_{\mathrm{z}} \end{pmatrix} - \mu \begin{pmatrix} \sigma^2\boldsymbol{C}_{\mathrm{z}}^T\boldsymbol{C}_{\mathrm{z}} & \boldsymbol{0} \\ \boldsymbol{0} & -\boldsymbol{I} \end{pmatrix} - \tilde{\boldsymbol{Q}} > 0 \tag{5.42}$$

holds with the LMI variables $\boldsymbol{P} \in \mathbb{R}^{(n+m+p)\times(n+m+p)}$ *symmetric and positive definite and the scalar* $\mu \in \mathbb{R}_0^+$. *Furthermore, the closed-loop cost function* (5.39) *is bounded.*

PROOF. Assume that there exists a matrix $\boldsymbol{P} > \boldsymbol{0}$ such that the LMI condition (5.42) is feasible. Pre-/post-multiplying (5.42) by the vector $\begin{pmatrix} \boldsymbol{z}^T(k) & \boldsymbol{e}_{\mathrm{y}}^T(k) \end{pmatrix}$ and its transpose, respectively, yields

$$\Delta V(k) < -\mu \begin{pmatrix} \boldsymbol{z}(k) \\ \boldsymbol{e}_{\mathrm{y}}(k) \end{pmatrix}^T \begin{pmatrix} \sigma^2\boldsymbol{C}_{\mathrm{z}}^T\boldsymbol{C}_{\mathrm{z}} & \boldsymbol{0} \\ \boldsymbol{0} & -\boldsymbol{I} \end{pmatrix} \begin{pmatrix} \boldsymbol{z}(k) \\ \boldsymbol{e}_{\mathrm{y}}(k) \end{pmatrix} - \begin{pmatrix} \boldsymbol{z}(k) \\ \boldsymbol{e}_{\mathrm{y}}(k) \end{pmatrix}^T \tilde{\boldsymbol{Q}} \begin{pmatrix} \boldsymbol{z}(k) \\ \boldsymbol{e}_{\mathrm{y}}(k) \end{pmatrix}. \tag{5.43}$$

Considering the definition of $\hat{\boldsymbol{y}}(k)$ and $\hat{\boldsymbol{x}}_{\mathrm{I}}(k)$ (5.4) and the event-triggering condition (5.12) a bound on the error variable $\boldsymbol{e}_{\mathrm{y}}(k)$ defined by (5.37) can be given as

$$\boldsymbol{e}_{\mathrm{y}}^T(k)\boldsymbol{e}_{\mathrm{y}}(k) \leq \sigma^2\boldsymbol{z}^T(k)\boldsymbol{C}_{\mathrm{z}}^T\boldsymbol{C}_{\mathrm{z}}\boldsymbol{z}(k). \tag{5.44}$$

Based on the bound (5.44) and the constraints $\mu \geq 0$ and $\tilde{\boldsymbol{Q}} \geq \boldsymbol{0}$, the right-hand side of inequality (5.43) is not larger than zero, guaranteeing the negative definiteness of $\Delta V(k)$, which implies that (5.38) is GAS. Due to the bound (5.22) it can be concluded from (5.43) that

$$\Delta V(k) < -\begin{pmatrix} \boldsymbol{z}(k) \\ \boldsymbol{e}_{\mathrm{y}}(k) \end{pmatrix}^T \tilde{\boldsymbol{Q}} \begin{pmatrix} \boldsymbol{z}(k) \\ \boldsymbol{e}_{\mathrm{y}}(k) \end{pmatrix} \tag{5.45}$$

always holds. Summing up (5.45) over $k = 0, ..., \infty$ yields

$$V(0) - \lim_{k\to\infty} V(k) > J. \tag{5.46}$$

As the closed-loop system (5.38) is GAS, $V(k) \to 0$ for $k \to \infty$. Thus, inequality (5.46) implies an upper bound

$$J < \boldsymbol{z}^T(0)\boldsymbol{P}\boldsymbol{z}(0) < \mathrm{tr}(\boldsymbol{P})\,\|\boldsymbol{z}(0)\|_2^2\,. \tag{5.47}$$

This completes the proof. □

The minimum of the upper bound given in (5.47) can be determined by applying the following Corollary.

Corollary 5.2 *Under event-triggered PI control* (5.3), (5.4), (5.12) *with given parameters* $\boldsymbol{K}$ *and* σ *a minimum upper bound of the cost function* (5.39) *is obtained from the*

LMI optimization problem

$$\min_{\boldsymbol{P}} \operatorname{tr}(\boldsymbol{P}) \quad \text{subject to} \tag{5.48a}$$

$$\begin{pmatrix} \boldsymbol{P} - \tilde{\boldsymbol{A}}_{\mathrm{z}}^T \boldsymbol{P} \tilde{\boldsymbol{A}}_{\mathrm{z}} & * \\ -\tilde{\boldsymbol{B}}_{\mathrm{z}}^T \boldsymbol{P} \tilde{\boldsymbol{A}}_{\mathrm{z}} & -\tilde{\boldsymbol{B}}_{\mathrm{z}}^T \boldsymbol{P} \tilde{\boldsymbol{B}}_{\mathrm{z}} \end{pmatrix} - \mu \begin{pmatrix} \sigma^2 \boldsymbol{C}_{\mathrm{z}}^T \boldsymbol{C}_{\mathrm{z}} & 0 \\ 0 & -\boldsymbol{I} \end{pmatrix} - \tilde{\boldsymbol{Q}} > 0 \tag{5.48b}$$

with the LMI variables $\boldsymbol{P} \in \mathbb{R}^{(n+m+p)\times(n+m+p)}$ symmetric and positive definite and the scalar $\mu \in \mathbb{R}_0^+$.

The theorem to determine the control parameters in a synthesized way is given as follows.

Theorem 5.4 *A solution to Problem 5.1 with respect to the event-triggering condition* (5.12) *is obtained from the LMI optimization problem*

$$\min_{\boldsymbol{Z},\boldsymbol{K}} \operatorname{tr}\left(\boldsymbol{Z}^{-1}\right) \quad \text{subject to} \tag{5.49a}$$

$$\boldsymbol{V}\boldsymbol{C}_{\mathrm{z}} = \boldsymbol{C}_{\mathrm{z}}\boldsymbol{G} \tag{5.49b}$$

$$\begin{pmatrix} \boldsymbol{G}^T + \boldsymbol{G} - \boldsymbol{Z} & * & * & * & * \\ 0 & \boldsymbol{V}^T + \boldsymbol{V} - \kappa \boldsymbol{I} & * & * & * \\ \boldsymbol{A}_{\mathrm{z}}\boldsymbol{G} + \boldsymbol{B}_{\mathrm{z}}\boldsymbol{U}\boldsymbol{C}_{\mathrm{z}} & \boldsymbol{B}_{\mathrm{z}}\boldsymbol{U} & \boldsymbol{Z} & * & * \\ \sigma \boldsymbol{C}_{\mathrm{z}}\boldsymbol{G} & 0 & 0 & \kappa \boldsymbol{I} & * \\ \boldsymbol{Q}^{1/2}\begin{pmatrix} \boldsymbol{G} \\ \boldsymbol{U}\boldsymbol{C}_{\mathrm{z}} \end{pmatrix} & \boldsymbol{Q}^{1/2}\begin{pmatrix} 0 \\ \boldsymbol{U} \end{pmatrix} & 0 & 0 & \boldsymbol{I} \end{pmatrix} > 0 \tag{5.49c}$$

with the LMI variables $\boldsymbol{G} \in \mathbb{R}^{(n+m+p)\times(n+m+p)}$, $\boldsymbol{V} \in \mathbb{R}^{2p\times 2p}$ and $\boldsymbol{U} \in \mathbb{R}^{m\times 2p}$ unrestricted, $\boldsymbol{Z} = \boldsymbol{P}^{-1}$ symmetric positive definite, and the positive scalar $\kappa \in \mathbb{R}^+$. The control gain results from

$$\boldsymbol{K} = \boldsymbol{U}\boldsymbol{V}^{-1}. \tag{5.50}$$

Proof. Assume that the conditions (5.49a) and (5.49b) are satisfied and consider further that $\boldsymbol{C}_{\mathrm{z}}$ is of full-row rank. Then (5.49b) implies that $\boldsymbol{V}$ is of full rank and thus invertible [DRI02]. Combining (5.49b) and (5.50) leads to

$$\boldsymbol{U}\boldsymbol{C}_{\mathrm{z}} = \boldsymbol{K}\boldsymbol{V}\boldsymbol{C}_{\mathrm{z}} = \boldsymbol{K}\boldsymbol{C}_{\mathrm{z}}\boldsymbol{G}. \tag{5.51}$$

Analogous to the derivation of (5.31) it can be shown that

$$\boldsymbol{V}^T\left(\kappa^{-1}\boldsymbol{I}\right)\boldsymbol{V} \geq \boldsymbol{V}^T + \boldsymbol{V} - \kappa \boldsymbol{I} \tag{5.52}$$

holds. Considering (5.31), (5.50), (5.51) and (5.52) condition (5.49c) implies

$$\begin{pmatrix} \boldsymbol{G}^T\boldsymbol{Z}^{-1}\boldsymbol{G} & * & * & * & * \\ 0 & \boldsymbol{V}^T\left(\kappa^{-1}\boldsymbol{I}\right)\boldsymbol{V} & * & * & * \\ \boldsymbol{A}_{\mathrm{z}}\boldsymbol{G} + \boldsymbol{B}_{\mathrm{z}}\boldsymbol{K}\boldsymbol{C}_{\mathrm{z}}\boldsymbol{G} & \boldsymbol{B}_{\mathrm{z}}\boldsymbol{K}\boldsymbol{V} & \boldsymbol{Z} & * & * \\ \sigma \boldsymbol{C}_{\mathrm{z}}\boldsymbol{G} & 0 & 0 & \kappa \boldsymbol{I} & * \\ \boldsymbol{Q}^{1/2}\begin{pmatrix} \boldsymbol{G} \\ \boldsymbol{K}\boldsymbol{C}_{\mathrm{z}}\boldsymbol{G} \end{pmatrix} & \boldsymbol{Q}^{1/2}\begin{pmatrix} 0 \\ \boldsymbol{K}\boldsymbol{V} \end{pmatrix} & 0 & 0 & \boldsymbol{I} \end{pmatrix} > 0. \tag{5.53}$$

Pre-/post-multiplying (5.53) by $\text{diag}(\boldsymbol{G}^{-T}, \boldsymbol{V}^{-T}, \boldsymbol{I}, \boldsymbol{I}, \boldsymbol{I})$ and $\text{diag}(\boldsymbol{G}^{-1}, \boldsymbol{V}^{-1}, \boldsymbol{I}, \boldsymbol{I}, \boldsymbol{I})$, respectively, yields

$$\begin{pmatrix} \boldsymbol{Z}^{-1} & * & * & * & * \\ 0 & \kappa^{-1}\boldsymbol{I} & * & * & * \\ \boldsymbol{A}_z + \boldsymbol{B}_z\boldsymbol{K}\boldsymbol{C}_z & \boldsymbol{B}_z\boldsymbol{K} & \boldsymbol{Z} & * & * \\ \sigma\boldsymbol{C}_z & 0 & 0 & \kappa\boldsymbol{I} & * \\ \boldsymbol{Q}^{1/2}\begin{pmatrix} \boldsymbol{I} \\ \boldsymbol{K}\boldsymbol{C}_z \end{pmatrix} & \boldsymbol{Q}^{1/2}\begin{pmatrix} 0 \\ \boldsymbol{K} \end{pmatrix} & 0 & 0 & \boldsymbol{I} \end{pmatrix} > 0. \tag{5.54}$$

Applying the Schur complement twice on (5.54) and substituting $\boldsymbol{P} = \boldsymbol{Z}^{-1}$ results in

$$\begin{pmatrix} \boldsymbol{P} - \tilde{\boldsymbol{A}}_z^T\boldsymbol{P}\tilde{\boldsymbol{A}}_z & * \\ -\tilde{\boldsymbol{B}}_z^T\boldsymbol{P}\tilde{\boldsymbol{A}}_z & -\tilde{\boldsymbol{B}}_z^T\boldsymbol{P}\tilde{\boldsymbol{B}}_z \end{pmatrix} - \kappa^{-1}\begin{pmatrix} \sigma^2\boldsymbol{C}_z^T\boldsymbol{C}_z & 0 \\ 0 & -\boldsymbol{I} \end{pmatrix} - \tilde{\boldsymbol{Q}} > 0. \tag{5.55}$$

Substituting $\mu = \kappa^{-1}$ in (5.55) it can be concluded from Theorem 5.3 that the closed-loop system (5.38) is GAS, and the cost function is upper bounded by $\text{tr}(\boldsymbol{P})\,\|\boldsymbol{z}(0)\|_2^2$. Based on the objective function (5.49a) the control gain $\boldsymbol{K}$ is determined such that this upper bound of the cost function is minimized. This completes the proof. □

5.3 Practical Stability Analysis

After designing the control parameters based on Theorem 5.2 or 5.4, global asymptotic stability is guaranteed, if the event-triggered control (5.3), (5.4), (5.11) or (5.12) is realized with $\epsilon = 0$. Otherwise, generally only practical stability can be achieved, i.e. the states do not converge to the origin but to a set containing the origin. This set is called region of practical stability in the following. In this section an approach is introduced to determine the region of practical stability $\mathcal{R}_{\text{PS}}$, which is positively invariant and globally attractive for the event-triggered controlled plant.

Definition 5.1 ([Bla99]) *A set $\mathcal{R}$ is said positively invariant for the system* (5.6) *controlled by the event-triggered control law* (5.3), (5.4), (5.11) *or* (5.12), *if $\boldsymbol{z}(0) \in \mathcal{R}$ implies $\boldsymbol{z}(k) \in \mathcal{R}$ for all $k \in \mathbb{N}$.*

Definition 5.2 ([RM09, Appendix B.2]) *A set $\mathcal{R}$ is said globally attractive for the system* (5.6) *controlled by the event-triggered control law* (5.3), (5.4), (5.11) *or* (5.12), *if each trajectory $\boldsymbol{z}(k)$ satisfies $\min_{\boldsymbol{z}^* \in \mathcal{R}} \|\boldsymbol{z}(k) - \boldsymbol{z}^*\|_2 \to 0$ as $k \to \infty$.*

In the following, the region of practical stability is described by an ellipsoidal set, which is represented as

$$\mathcal{R}_{\text{PS}} = \left\{\boldsymbol{z} \in \mathbb{R}^{n+m+p} : \boldsymbol{z}^T\boldsymbol{\Theta}\boldsymbol{z} \leq 1\right\} \tag{5.56}$$

where $\boldsymbol{\Theta} \in \mathbb{R}^{(n+m+p)\times(n+m+p)}$ is symmetric and positive definite. For given control and event-triggering parameters $\boldsymbol{K}$, σ and ϵ the aim is to determine a region of practical stability $\mathcal{R}_{\text{PS}}$. The set $\mathcal{R}_{\text{PS}}$ can then be interpreted as a guaranteed steady state performance.

5.3.1 Practical Stability Analysis for Control Input Based Event-Triggered PI Control

Under the control input based event-triggering condition (5.11) the practical stability can be analyzed based on the following Theorem 5.5, resulting in a positively invariant and globally attractive set, which describes the region of practical stability.

Theorem 5.5 *The set* $\mathcal{R}_{\text{PS}}$ *defined by* (5.56) *is positively invariant and globally attractive for the discrete-time closed-loop system* (5.14) *under event-triggered PI control* (5.3), (5.4), (5.11) *if the LMI optimization problem*

$$\min_{\boldsymbol{\Theta}} -\log\det(\boldsymbol{\Theta}) \quad \text{subject to} \tag{5.57a}$$

$$\begin{pmatrix} (1-\mu_2)\boldsymbol{\Theta} - \tilde{\boldsymbol{A}}_{\text{z}}^T\boldsymbol{\Theta}\tilde{\boldsymbol{A}}_{\text{z}} - \mu_1\sigma^2\boldsymbol{C}_{\text{z}}^T\boldsymbol{K}^T\boldsymbol{K}\boldsymbol{C}_{\text{z}} & * & * \\ -\boldsymbol{B}_{\text{z}}^T\boldsymbol{\Theta}\tilde{\boldsymbol{A}}_{\text{z}} & -\boldsymbol{B}_{\text{z}}^T\boldsymbol{\Theta}\boldsymbol{B}_{\text{z}} + \mu_1\boldsymbol{I} & * \\ \boldsymbol{0} & \boldsymbol{0} & -\mu_1\epsilon + \mu_2 \end{pmatrix} > 0 \tag{5.57b}$$

with the LMI variables $\boldsymbol{\Theta} \in \mathbb{R}^{(n+m+p)\times(n+m+p)}$ *symmetric and positive definite and a nonnegative scalar* $\mu_1 \in \mathbb{R}_0^+$ *obtains a solution. The fixed scalar variable* $\mu_2 \in (0,1)$ *is predefined by gridding.*

Proof. The proof is divided into two parts. The first part focuses on the global attraction of the set $\mathcal{R}_{\text{PS}}$ while the second part focuses on its positive invariance.

For proving the global attraction the Lyapunov function

$$W(k) = \boldsymbol{z}^T(k)\boldsymbol{\Theta}\boldsymbol{z}(k) \tag{5.58}$$

is defined. Assume that there exists a matrix $\boldsymbol{\Theta}$ such that condition (5.57b) is satisfied. Pre-/post-multiplying (5.57b) by the vector $\begin{pmatrix}\boldsymbol{z}^T(k) & \boldsymbol{e}_{\text{u}}^T(k) & 1\end{pmatrix}$ and its transpose, respectively, yields

$$\begin{aligned} \Delta W(k) < &-\mu_1 \begin{pmatrix} \boldsymbol{z}(k) \\ \boldsymbol{e}_{\text{u}}(k) \\ 1 \end{pmatrix}^T \begin{pmatrix} \sigma^2\boldsymbol{C}_{\text{z}}^T\boldsymbol{K}^T\boldsymbol{K}\boldsymbol{C}_{\text{z}} & * & * \\ \boldsymbol{0} & -\boldsymbol{I} & * \\ \boldsymbol{0} & \boldsymbol{0} & \epsilon \end{pmatrix} \begin{pmatrix} \boldsymbol{z}(k) \\ \boldsymbol{e}_{\text{u}}(k) \\ 1 \end{pmatrix} \\ &-\mu_2 \begin{pmatrix} \boldsymbol{z}(k) \\ \boldsymbol{e}_{\text{u}}(k) \\ 1 \end{pmatrix}^T \begin{pmatrix} \boldsymbol{\Theta} & * & * \\ \boldsymbol{0} & \boldsymbol{0} & * \\ \boldsymbol{0} & \boldsymbol{0} & -1 \end{pmatrix} \begin{pmatrix} \boldsymbol{z}(k) \\ \boldsymbol{e}_{\text{u}}(k) \\ 1 \end{pmatrix} \end{aligned} \tag{5.59}$$

where $\Delta W(k) = W(k+1) - W(k)$ is the difference of the Lyapunov function along the trajectories of the closed-loop system (5.14). Including the absolute threshold $\epsilon \neq 0$, the bound on the error variable $\boldsymbol{e}_\mathrm{u}(k)$ (5.22) is given by

$$\boldsymbol{e}_\mathrm{u}^T(k)\boldsymbol{e}_\mathrm{u}(k) \leq \sigma^2 \boldsymbol{z}^T(k)\boldsymbol{C}_\mathrm{z}^T\boldsymbol{K}^T\boldsymbol{K}\boldsymbol{C}_\mathrm{z}\boldsymbol{z}(k) + \epsilon \tag{5.60}$$

which can equivalently be written as

$$\begin{pmatrix} \boldsymbol{z}(k) \\ \boldsymbol{e}_\mathrm{u}(k) \\ 1 \end{pmatrix}^T \begin{pmatrix} \sigma^2\boldsymbol{C}_\mathrm{z}^T\boldsymbol{K}^T\boldsymbol{K}\boldsymbol{C}_\mathrm{z} & * & * \\ 0 & -\boldsymbol{I} & * \\ 0 & 0 & \epsilon \end{pmatrix} \begin{pmatrix} \boldsymbol{z}(k) \\ \boldsymbol{e}_\mathrm{u}(k) \\ 1 \end{pmatrix} \geq 0. \tag{5.61}$$

Moreover, $\boldsymbol{z}(k) \notin \mathcal{R}_\mathrm{PS}$ is equal to

$$\begin{pmatrix} \boldsymbol{z}(k) \\ \boldsymbol{e}_\mathrm{u}(k) \\ 1 \end{pmatrix}^T \begin{pmatrix} \boldsymbol{\Theta} & * & * \\ 0 & 0 & * \\ 0 & 0 & -1 \end{pmatrix} \begin{pmatrix} \boldsymbol{z}(k) \\ \boldsymbol{e}_\mathrm{u}(k) \\ 1 \end{pmatrix} > 0. \tag{5.62}$$

As condition (5.61) is always satisfied due to the bound on the error variable $\boldsymbol{e}_\mathrm{u}(k)$, it can then be concluded that the right-hand of (5.59) is smaller than zero for all $\boldsymbol{z}(k) \notin \mathcal{R}_\mathrm{PS}$ as $\mu_1 \geq 0$ and $\mu_2 > 0$. Consequently, the Lyapunov function is strictly decreasing along solutions of the closed-loop system as long as the set $\mathcal{R}_\mathrm{PS}$ is not reached, showing the convergence of the trajectories to set $\mathcal{R}_\mathrm{PS}$.

After proving the global attraction, also the positive invariance of the set $\mathcal{R}_\mathrm{PS}$ can be shown based on condition (5.57b). Therefore, it is to show that $\boldsymbol{z}(k) \in \mathcal{R}_\mathrm{PS}$ implies $\boldsymbol{z}(k+1) \in \mathcal{R}_\mathrm{PS}$. Substituting $\kappa_2 = 1 - \mu_2$, condition (5.57b) is equivalent to

$$\begin{pmatrix} \kappa_2\boldsymbol{\Theta} - \tilde{\boldsymbol{A}}_\mathrm{z}^T\boldsymbol{\Theta}\tilde{\boldsymbol{A}}_\mathrm{z} - \mu_1\sigma^2\boldsymbol{C}_\mathrm{z}^T\boldsymbol{K}^T\boldsymbol{K}\boldsymbol{C}_\mathrm{z} & * & * \\ -\boldsymbol{B}_\mathrm{z}^T\boldsymbol{\Theta}\tilde{\boldsymbol{A}}_\mathrm{z} & -\boldsymbol{B}_\mathrm{z}^T\boldsymbol{\Theta}\boldsymbol{B}_\mathrm{z} + \mu_1\boldsymbol{I} & * \\ 0 & 0 & -\mu_1\epsilon + 1 - \kappa_2 \end{pmatrix} > 0 \tag{5.63}$$

where $\kappa_2 \in (0,1)$. Pre-/post-multiplying (5.63) by the vector $\begin{pmatrix} \boldsymbol{z}^T(k) & \boldsymbol{e}_\mathrm{u}^T(k) & 1 \end{pmatrix}$ and its transpose, respectively, results in

$$\begin{aligned} &\begin{pmatrix} \boldsymbol{z}(k) \\ \boldsymbol{e}_\mathrm{u}(k) \\ 1 \end{pmatrix}^T \begin{pmatrix} -\tilde{\boldsymbol{A}}_\mathrm{z}^T\boldsymbol{\Theta}\tilde{\boldsymbol{A}}_\mathrm{z} & * & * \\ -\boldsymbol{B}_\mathrm{z}^T\boldsymbol{\Theta}\tilde{\boldsymbol{A}}_\mathrm{z} & -\boldsymbol{B}_\mathrm{z}^T\boldsymbol{\Theta}\boldsymbol{B}_\mathrm{z} & * \\ 0 & 0 & 1 \end{pmatrix} \begin{pmatrix} \boldsymbol{z}(k) \\ \boldsymbol{e}_\mathrm{u}(k) \\ 1 \end{pmatrix} > \\ &\mu_1 \begin{pmatrix} \boldsymbol{z}(k) \\ \boldsymbol{e}_\mathrm{u}(k) \\ 1 \end{pmatrix}^T \begin{pmatrix} \sigma^2\boldsymbol{C}_\mathrm{z}^T\boldsymbol{K}^T\boldsymbol{K}\boldsymbol{C}_\mathrm{z} & * & * \\ 0 & -\boldsymbol{I} & * \\ 0 & 0 & \epsilon \end{pmatrix} \begin{pmatrix} \boldsymbol{z}(k) \\ \boldsymbol{e}_\mathrm{u}(k) \\ 1 \end{pmatrix} + \kappa_2 \begin{pmatrix} \boldsymbol{z}(k) \\ \boldsymbol{e}_\mathrm{u}(k) \\ 1 \end{pmatrix}^T \begin{pmatrix} -\boldsymbol{\Theta} & * & * \\ 0 & 0 & * \\ 0 & 0 & 1 \end{pmatrix} \begin{pmatrix} \boldsymbol{z}(k) \\ \boldsymbol{e}_\mathrm{u}(k) \\ 1 \end{pmatrix}. \end{aligned} \tag{5.64}$$

Furthermore, $\boldsymbol{z}(k) \in \mathcal{R}_\mathrm{PS}$ is equivalent to

$$\begin{pmatrix} \boldsymbol{z}(k) \\ \boldsymbol{e}_\mathrm{u}(k) \\ 1 \end{pmatrix}^T \begin{pmatrix} -\boldsymbol{\Theta} & * & * \\ 0 & 0 & * \\ 0 & 0 & 1 \end{pmatrix} \begin{pmatrix} \boldsymbol{z}(k) \\ \boldsymbol{e}_\mathrm{u}(k) \\ 1 \end{pmatrix} > 0. \tag{5.65}$$

As (5.61) holds and $z(k) \in \mathcal{R}_{\mathrm{PS}}$, the right-hand side of (5.64) is nonnegative with the scalars $\mu_2 \geq 0$ and $\kappa_2 > 0$, leading to

$$\begin{pmatrix} z(k) \\ e_{\mathrm{u}}(k) \\ 1 \end{pmatrix}^T \begin{pmatrix} -\tilde{A}_{\mathrm{z}}^T \Theta \tilde{A}_{\mathrm{z}} & * & * \\ -B_{\mathrm{z}}^T \Theta \tilde{A}_{\mathrm{z}} & -B_{\mathrm{z}}^T \Theta B_{\mathrm{z}} & * \\ 0 & 0 & 1 \end{pmatrix} \begin{pmatrix} z(k) \\ e_{\mathrm{u}}(k) \\ 1 \end{pmatrix} > 0. \tag{5.66}$$

Based on the closed-loop system (5.14), inequality (5.66) equals $z^T(k+1)\Theta z(k+1) < 1$, i.e. $z(k+1) \in \mathcal{R}_{\mathrm{PS}}$. By induction it follows that $z(k+l) \in \mathcal{R}_{\mathrm{PS}}$ for all $l \in \mathbb{N}$. This completes the proof. □

The objective function (5.57a) in the LMI optimization is selected such that a small region of practical stability is achieved. As the volume of an ellipsoid defined as (5.56) is proportional to $(\det \Theta)^{-1/2}$, minimizing $-\log \det(\Theta)$ corresponds to minimizing the volume of the ellipsoid [BEFB94, Section 2.2.4]. In general, also other size criteria can be used for such problems, see [TGGQ11, Section 2.2.5.1] for a discussion on different size criteria.

It is worth to note that the result of the optimization problem (5.57) only gives an estimate of the region of practical stability based on a sufficient condition. The real size of the region and thus the steady state performance is affected by the parameter ϵ. By decreasing ϵ the steady state output error $y(t) - r$ can be made arbitrary small, which may however increase the number of triggered events, as can be seen in the simulation results in Section 5.4 and in the experimental studies in Section 5.5.

5.3.2 Practical Stability Analysis for Output Based Event-Triggered PI Control

In a similar manner as in Section 5.3.1 practical stability can be analyzed under the output based event-triggering condition (5.12). Equivalently, a theorem is proposed, resulting in a positively invariant and globally attractive set, which describes the region of practical stability.

Theorem 5.6 *The set* $\mathcal{R}_{\mathrm{PS}}$ *defined by* (5.56) *is positively invariant and globally attractive for the discrete-time closed-loop system* (5.38) *under event-triggered PI control* (5.3), (5.4), (5.12) *if the LMI optimization problem*

$$\min_{\Theta} -\log\det(\Theta) \quad \text{subject to} \tag{5.67a}$$

$$\begin{pmatrix} (1-\mu_2)\Theta - \tilde{A}_{\mathrm{z}}^T \Theta \tilde{A}_{\mathrm{z}} - \mu_1 \sigma^2 C_{\mathrm{z}}^T C_{\mathrm{z}} & * & * \\ -\tilde{B}_{\mathrm{z}}^T \Theta \tilde{A}_{\mathrm{z}} & -\tilde{B}_{\mathrm{z}}^T \Theta \tilde{B}_{\mathrm{z}} + \mu_1 I & * \\ 0 & 0 & -\mu_1 \epsilon + \mu_2 \end{pmatrix} > 0 \tag{5.67b}$$

with the LMI variables $\boldsymbol{\Theta} \in \mathbb{R}^{(n+m+p)\times(n+m+p)}$ *symmetric and positive definite and a nonnegative scalar* $\mu_1 \in \mathbb{R}_0^+$ *obtains a solution. The fixed scalar variable* $\mu_2 \in (0,1)$ *is predefined by gridding.*

PROOF. The proof follows the same lines as the one of Theorem (5.5) taking into account that the bound on the error variable $\boldsymbol{e}_{\mathrm{y}}(k)$ is given by

$$\boldsymbol{e}_{\mathrm{y}}^T(k)\boldsymbol{e}_{\mathrm{y}}(k) \leq \sigma^2 \boldsymbol{z}^T(k)\boldsymbol{C}_{\mathrm{z}}^T\boldsymbol{C}_{\mathrm{z}}\boldsymbol{z}(k) + \epsilon \tag{5.68}$$

and the closed-loop system follows (5.38). □

5.4 Illustrative Examples and Comparisons

For evaluating the effectiveness of the proposed event-triggered PI control approach, the method is applied to two illustrative examples. It is analyzed how the event-triggering parameters σ and ϵ as well as the control gain $\boldsymbol{K}$, resulting from the control synthesis, influence the control performance and the transmission rate. The transmission rate is evaluated by the number of triggered events within the simulation time. The simulations are realized with MATLAB/Simulink.

5.4.1 Related Event-Triggered PI Control Approaches

Further, a comparison with event-triggered PI control methods with an absolute threshold policy is conducted. One method is the *Simple event-triggered PI control* method [Årz99], where the event-triggering condition is verified periodically under a fixed period length. In the other method the event-triggering condition is monitored continuously [LKJ12] and is therefore denoted as *Continuous-time event-triggered PI control.* In the following the two approaches for comparison are elaborated.

Simple Event-Triggered PI Control

The simple event-triggered PI control strategy is formalized as follows. Analogous to the proposed method, the output vector is sampled periodically at every time instant t_k, with $t_{k+1} = t_k + h$, and forwarded to the event generator. If an event is triggered, the control input is computed by the PI control law given by

$$\boldsymbol{x}_{\mathrm{I}}(k+1) = \boldsymbol{x}_{\mathrm{I}}(k) + h_{\mathrm{act}}(k)\big(\hat{\boldsymbol{y}}(k) - \boldsymbol{r}\big) \tag{5.69a}$$

$$\boldsymbol{u}(k) = \boldsymbol{K}_{\mathrm{I}}\boldsymbol{x}_{\mathrm{I}}(k) + \boldsymbol{K}_{\mathrm{P}}\big(\hat{\boldsymbol{y}}(k) - \boldsymbol{r}\big) \tag{5.69b}$$

where $\hat{\boldsymbol{y}}(k) = \boldsymbol{y}(k)$ and $h_{\text{act}}(k)$ is the elapsed time since the previous event occurrence, otherwise the variables are held, i.e.

$$\boldsymbol{x}_{\text{I}}(k+1) = \boldsymbol{x}_{\text{I}}(k) \tag{5.70a}$$

$$\boldsymbol{u}(k) = \boldsymbol{u}(k-1) \tag{5.70b}$$

and $\hat{\boldsymbol{y}}(k) = \hat{\boldsymbol{y}}(k-1)$. An event is triggered, if the difference between the current value of the output vector $\boldsymbol{y}(k)$ and the value of the output vector when an event was triggered last time $\hat{\boldsymbol{y}}(k-1)$ is greater than an absolute threshold ϵ_{DT}, or if the elapsed time since the last event $h_{\text{act}}(k)$ exceeds a limit h_{max}. Thus, the event-triggering condition is given by

$$\|\hat{\boldsymbol{y}}(k-1) - \boldsymbol{y}(k)\|_2 > \epsilon_{\text{DT}} \quad \text{or} \quad h_{\text{act}}(k) \geq h_{\text{max}}. \tag{5.71}$$

In this approach, one aim is to keep a low computational complexity. Opposite to (5.3) the integration of the output error is not calculated periodically but only when an event occurs. Additionally, the integrator state is not required in the event generator such that integrator state can be calculated in the control task T_{PI} making the transmission of the integrator state obsolete, see Figure 5.1.

Continuous-Time Event-Triggered PI Control

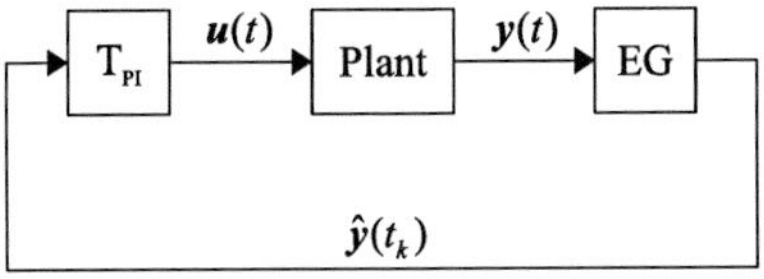

Figure 5.3: Continuous-time event-triggered control architecture

Besides, a continuous-time event-triggered PI control strategy is taken into consideration, where the output is monitored continuously [OMT02, LJ12]. In this approach, the control input $\boldsymbol{u}(t)$ is computed continuously based on the PI control law given by

$$\dot{\boldsymbol{x}}_{\text{I}}(t) = \hat{\boldsymbol{y}}(t_k) - \boldsymbol{r} \tag{5.72a}$$

$$\boldsymbol{u}(t) = \boldsymbol{K}_{\text{I}}\boldsymbol{x}_{\text{I}}(t) + \boldsymbol{K}_{\text{P}}\big(\hat{\boldsymbol{y}}(t_k) - \boldsymbol{r}\big), \tag{5.72b}$$

for $t \in [t_k, t_{k+1})$, where $\hat{\boldsymbol{y}}(t_k)$ is the last transmitted value of the output vector. The controller receives new information about the output vector only at time instants t_k when an event is invoked. Therefore, the output vector $\boldsymbol{y}(t)$ is continuously sampled and forwarded to the event generator, where an event is triggered at the time instant $t = t_{k+1}$ whenever the event triggering condition

$$\|\boldsymbol{y}(t) - \hat{\boldsymbol{y}}(t_k)\|_2 = \epsilon_{\text{CT}} \tag{5.73}$$

is satisfied. The corresponding structure is depicted in Figure 5.3. Additionally, the continuous-time approach requires the determination of a minimum inter-event time, which gives a lower bound on the elapsed time between two consecutive events. If the inter-event time becomes arbitrary small the completion of the execution of the control task and the communication cannot be realized. This so-called Zeno-behavior must be excluded, which is discussed in [LJ12, LKJ12].

5.4.2 Evaluation

For evaluating the different event-triggered PI control strategies two measures are taken into account, considering the transmission effort and control performance. The transmission effort is quantified by the number of triggered events N_{event}, i.e. the number of transmitted packets, within a defined evaluation time T_{eval}, which corresponds to the simulation time. The control performance is measured by the evaluation cost function

$$J_{\text{eval}} = \int_0^{T_{\text{eval}}} \begin{pmatrix} \tilde{\boldsymbol{x}}_{\text{p}}(t) \\ \tilde{\boldsymbol{u}}(t-\tau) \end{pmatrix}^T \begin{pmatrix} \boldsymbol{Q}_{\text{c}} & \boldsymbol{0} \\ \boldsymbol{0} & \boldsymbol{R}_{\text{c}} \end{pmatrix} \begin{pmatrix} \tilde{\boldsymbol{x}}_{\text{p}}(t) \\ \tilde{\boldsymbol{u}}(t-\tau) \end{pmatrix} dt. \tag{5.74}$$

This corresponds to the cost function (5.7) with $\boldsymbol{Q}_{\text{I}} = \boldsymbol{0}$. There are two reasons for neglecting the integrator state $\boldsymbol{x}_{\text{I}}(k)$ in the evaluation cost function J_{eval}. First, the state vector $\boldsymbol{x}_{\text{I}}(k)$ does not give a real indication on the control performance. Second, the cost function J is evaluated with respect to the deviation variables (5.8). The deviation variables are defined based on the steady state values, which are computed by (5.9). The steady state values of $\boldsymbol{x}_{\text{p}}(t)$ and $\boldsymbol{u}(t)$ can also be calculated as

$$\begin{pmatrix} \boldsymbol{x}_{\text{p}}^{\text{ss}} \\ \boldsymbol{u}^{\text{ss}} \end{pmatrix} = \begin{pmatrix} \boldsymbol{A}_{\text{p}} & \boldsymbol{B}_{\text{p}} \\ \boldsymbol{C}_{\text{p}} & \boldsymbol{0} \end{pmatrix}^{-1} \begin{pmatrix} \boldsymbol{0} \\ \boldsymbol{r} \end{pmatrix}. \tag{5.75}$$

Consequently, the steady state $\boldsymbol{x}_{\text{I}}^{\text{ss}}$ depends on the control parameter $\boldsymbol{K}_{\text{I}}$. Under the assumption that the original initial states are all zero, i.e. $\boldsymbol{x}_{\text{p}}(0) = \boldsymbol{0}$, $\boldsymbol{u}(0) = \boldsymbol{0}$ and $\boldsymbol{x}_{\text{I}}(0) = \boldsymbol{0}$, and a reference signal $\boldsymbol{r}$ shall be tracked, this means that the initial value of the deviation variable $\tilde{\boldsymbol{x}}_{\text{I}}(0) = \boldsymbol{0} - \boldsymbol{x}_{\text{I}}^{\text{ss}}$ depends on the control parameter $\boldsymbol{K}_{\text{I}}$. As each method usually works with different control parameters, the initial states of the deviation variables $\tilde{\boldsymbol{x}}_{\text{I}}(t)$ may differ, such that the cost function J cannot work as a good performance measure. On the other hand, the integrator state is included in the cost function J for the control synthesis, as its consideration gives more design possibilities for realizing a controller according to the design requirements. Simulations have shown that neglecting the integrator state in the cost function for the control synthesis can lead to deficient results.

Example 5.1 Consider the example taken from [LKJ12] with a plant (5.1) described by the continuous-time state equation

$$\begin{aligned} \dot{x}_{\text{p}}(t) &= 0.1 x_{\text{p}}(t) + u(t-\tau) \\ y(t) &= x_{\text{p}}(t) \end{aligned} \tag{5.76}$$

with the initial state $x_p(0) = 0$. The delay, which measures the communication time, is assumed $\tau = 5\,\text{ms}$ for each method, and the parameters of the continuous-time cost function are selected as $Q_c = 2.5$, $Q_I = 2.0$ and $R_c = 0.5$. For applying the proposed event-triggered PI control strategy, the continuous-time state equation (5.76) and continuous-time cost function (5.7) are discretized with the step size $h = 10\,\text{ms}$. In the following simulation, a reference signal $r = 1$ is tracked and the simulation time is $T_{\text{eval}} = 10\,\text{s}$.

Control Input Based ETC

Based on the given parameters first the control input based approach (Section 5.2.1) is realized. Theorem 5.2 is implemented in MATLAB using the toolbox CVX [GB12, GB08] and the SeDuMi solver [Stu99] for several values of σ with $\epsilon = 0$. For the simulation of the event-triggered PI control under the control input based event, the initial values of

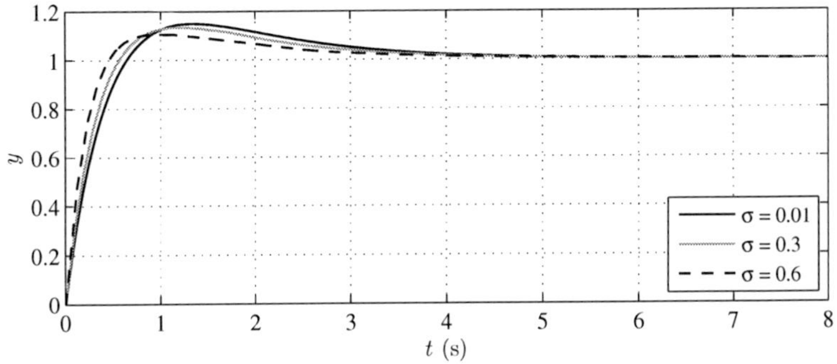

Figure 5.4: Output $y(t)$ for different values of σ (control input based ETC)

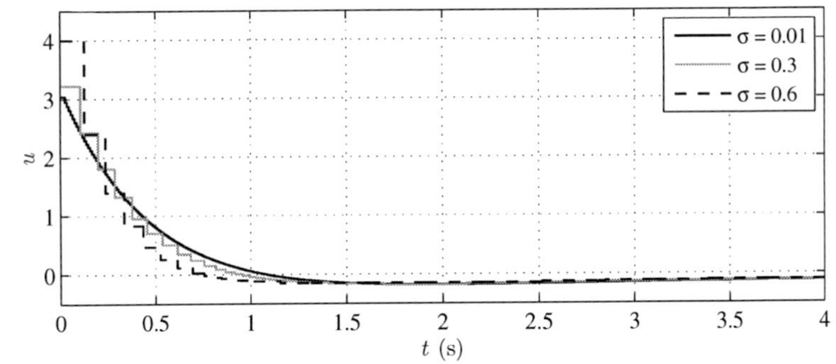

Figure 5.5: Input $u(t - \tau)$ for different values of σ (control input based ETC)

the controller and event generator states are set $x_\mathrm{I}(0) = 0$, $\hat{x}_\mathrm{I}(-1) = 0$ and $\hat{y}(-1) = r$, such that at the initial time instant $k = 0$ an event is triggered. For a representative set of values of σ, the performance indices are given in Table 5.1. The parameter ϵ is set zero. It shows that increasing the value of σ will reduce the number of triggered events, however, at the expense of control performance degradation and vice versa. In order to illustrate how the control performance degrades with increasing σ, an excerpt of the output and input behavior is shown in Figure 5.4 and Figure 5.5, respectively, for some values of σ. It can be seen in Figure 5.4 that for a higher value of σ the output converges faster to the reference signal with less overshooting, resulting in a better output behavior. On the other hand, the control input is larger and less smooth for $\sigma = 0.6$ than for the other cases, see also the control gain $\boldsymbol{K}$ in Table 5.1, leading overall to a higher cost. The example demonstrates how a balance between control performance and resource utilization can be achieved using the tuning parameter σ.

σ	J_eval	N_event	$\boldsymbol{K}$
0.01	1.467	568	$(-3.0403 \quad -1.9394)$
0.1	1.487	135	$(-3.0578 \quad -1.9480)$
0.3	1.558	56	$(-3.2179 \quad -2.0270)$
0.6	1.801	32	$(-3.9952 \quad -2.3928)$

Table 5.1: Evaluation under control input based ETC (Theorem 5.2)

Output Based ETC

In the following, the output based strategy (Section 5.2.2) is realized in simulation implementing Theorem 5.4 for a set of values of σ with $\epsilon = 0$. In the output based approach, the initial values of the controller and event generator states are set $x_\mathrm{I}(0) = 0$, $\hat{x}_\mathrm{I}(-1) = x_\mathrm{I}^\mathrm{ss}$ and $\hat{y}(-1) = r$.

Table 5.2 shows the performance indices of the simulation results, which are given in Figure 5.6 and Figure 5.7 for selected values of σ. The results illustrate the same trend as the results given in Table 5.1 for the other event-triggering condition (5.11). For small values of σ both strategies lead to similar results, however for increasing σ the output based strategy leads to a comparatively stronger degradation of the control

σ	J_eval	N_event	$\boldsymbol{K}$
0.01	1.468	544	$(-3.0488 \quad -1.9479)$
0.05	1.515	185	$(-3.2826 \quad -2.1791)$
0.1	1.797	118	$(-4.6062 \quad -3.4683)$
0.12	2.374	100	$(-6.7884 \quad -5.5533)$

Table 5.2: Evaluation under output based ETC (Theorem 5.4)

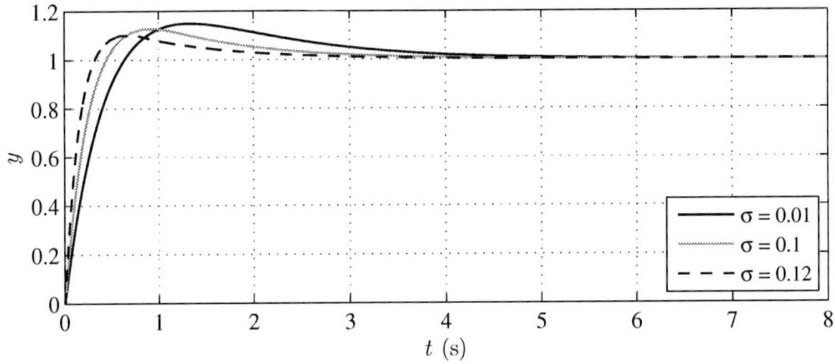

Figure 5.6: Output $y(t)$ for different values of σ (output based ETC)

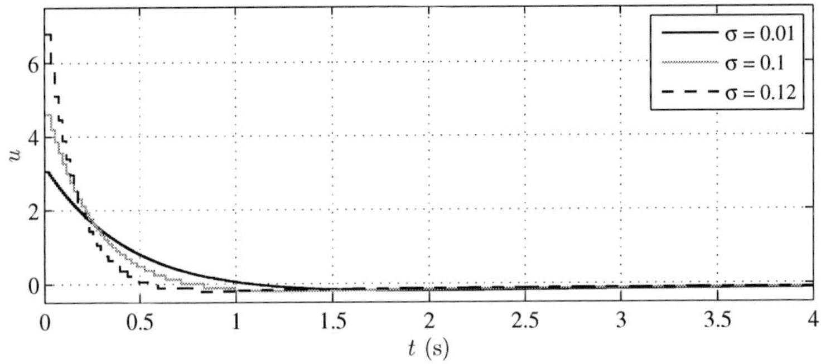

Figure 5.7: Input $u(t-\tau)$ for different values of σ (output based ETC)

performance. Under a reasonable increase of the cost function J_{eval}, only a limited reduction of triggered events can be realized in this example.

An improved resource utilization can be achieved by solving Corollary 5.2 for $\sigma > \sigma_0$ with a given control gain $\boldsymbol{K}$, which results from Theorem 5.4 for a given σ_0. This heuristic procedure is motivated by the fact that the extension of Theorem 5.3 towards the PI control synthesis in Theorem 5.4 induces some conservatism, e.g. (5.52), in order to formulate the control synthesis problem as an LMI optimization problem. The implicated reduction of conservatism by this heuristic procedure is illustrated in Figure 5.8, where the dashed line shows the upper bound $\text{tr}(\boldsymbol{P})$ as a function of σ based on Corollary 5.2 with always the same control gain $\boldsymbol{K} = \begin{pmatrix} -3.0488 & -1.9479 \end{pmatrix}$ resulting from Theorem 5.4 with $\sigma_0 = 0.01$. The solid grey line shows the upper bound $\text{tr}(\boldsymbol{P})$ as a function of σ based on Theorem 5.4, where the control gain $\boldsymbol{K}$ differs for each value of σ.

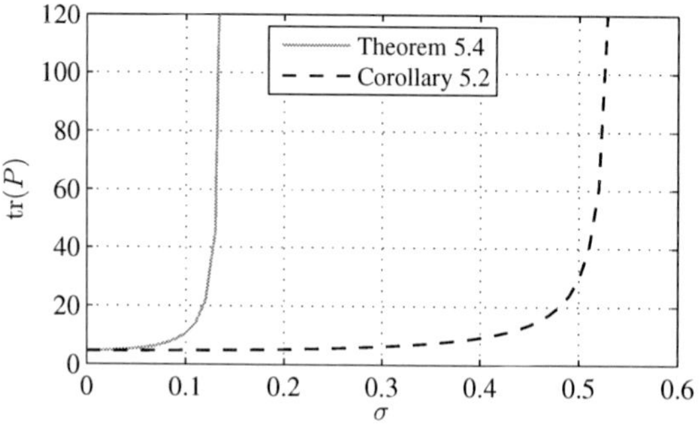

Figure 5.8: Upper bound $\mathrm{tr}(\boldsymbol{P})$ as a function of σ (output based ETC)

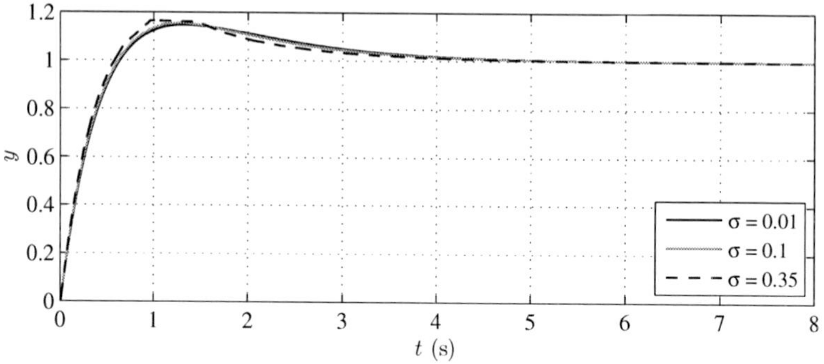

Figure 5.9: Output $y(t)$ for different values of σ (output based ETC, Corollary 5.2)

The simulation results for selected values of σ are given in Figure 5.9 and summarized in Table 5.3. The output behavior for $\sigma = 0.01$ and $\sigma = 0.1$ is very similar, whereas for $\sigma = 0.1$ less events are triggered. For $\sigma = 0.35$ the output behavior is less smooth than for the other cases, leading to a control performance degradation but also to a significant resource utilization reduction.

σ	J_{eval}	N_{event}	$\boldsymbol{K}$
0.01	1.468	544	$(-3.0488 \quad -1.9479)$
0.1	1.488	103	$(-3.0488 \quad -1.9479)$
0.35	1.547	35	$(-3.0488 \quad -1.9479)$

Table 5.3: Evaluation under output based ETC (Corollary 5.2)

Comparison

To complete the evaluation of this example a comparison with the two approaches presented in Section 5.4.1 is made. Thereby, for the control input based strategy the control parameters are design applying Theorem 5.2 with $\sigma = 0.1$. The event-triggering condition (5.11) is then realized in simulation with $\sigma = 0.1$ and $\epsilon = 3 \cdot 10^{-5}$. Practical stability is verified based on Theorem 5.5, which is implemented using the MATLAB toolbox YALMIP [Löf04] and solved using the SeDuMi solver [Stu99].

For the output based strategy the control synthesis (Theorem 5.4) is applied for $\sigma_0 = 0.01$ in order to determine the control parameters. The event-triggered implementation is then realized with $\sigma = 0.2$ and $\epsilon = 1 \cdot 10^{-6}$ after verifying practical stability with Theorem 5.6.

The event-triggered PI control strategies considered for comparison are so-called emulation-based approaches, see the discussion in Section 1.3.2 or [HDT13, Section IV]. Thus, the control parameters of the simple event-triggered PI control strategy are determined under the assumption of a time-triggered periodic implementation with the constant step size $h = 10\,\text{ms}$. As performance measure the cost function (5.10) with the same weighting matrices as indicated above is considered, see Appendix A.4.1. The event-triggering parameters are set $\epsilon_{\text{DT}} = 0.03$ and $h_{\text{max}} = 20h$. For the design of the control parameters of the continuous-time event-triggered PI control strategy, first a continuous-time implementation is assumed. After building up a continuous-time augmented system model with the state vector $\begin{pmatrix} \boldsymbol{x}_{\text{p}}^T(t) & \boldsymbol{x}_{\text{I}}^T(t) \end{pmatrix}^T$, the continuous-time LQR method is applied for designing the control parameters, see Appendix A.4.2. The event-triggering parameter is chosen as $\epsilon_{\text{CT}} = 0.02$.

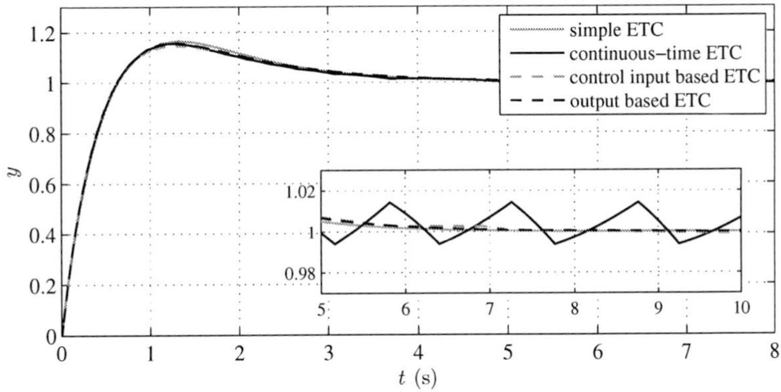

Figure 5.10: Output $y(t)$ under different event-triggered control strategies

Method	J_{eval}	N_{event}	$\boldsymbol{K}$
Simple ETC	1.485	74	$(-3.0402 \quad -1.9394)$
Continuous-time ETC	1.466	72	$(-3.1017 \quad -2.0)$
Control input based ETC	1.487	67	$(-3.0578 \quad -1.9480)$
Output based ETC	1.488	66	$(-3.0488 \quad -1.9479)$

Table 5.4: Evaluation under different event-triggered control strategies ($T_{\text{eval}} = 10\,\text{s}$)

Method	J_{eval}	N_{event}	$\boldsymbol{K}$
Simple ETC	1.485	274	$(-3.0402 \quad -1.9394)$
Continuous-time ETC	1.487	123	$(-3.1017 \quad -2.0)$
Control input based ETC	1.487	95	$(-3.0578 \quad -1.9480)$
Output based ETC	1.488	99	$(-3.0488 \quad -1.9479)$

Table 5.5: Evaluation under different event-triggered control strategies ($T_{\text{eval}} = 50\,\text{s}$)

The delay is set $\tau = 5\,\text{ms}$ for all approaches in the simulation.[1] Simulation results are given in Figure 5.10. From the transient output performance point of view, all strategies show a similar behavior, with the continuous-time approach leading to a smaller cost than the other approaches for $T_{\text{eval}} = 10\,\text{s}$. However, under the output based and control input based ETC approach less events are triggered, see Table 5.4. Considering the steady state performance, the continuous-time strategy leads to an oscillation of the output around the setpoint, zoomed in Figure 5.10, due to the use of the absolute threshold policy (5.73). Especially for a longer evaluation time, e.g. $T_{\text{eval}} = 50\,\text{s}$, the resulting control performance degradation due to the oscillation can be clearly seen by the increased cost, see the comparison between J_{eval} of the continuous-time approach in Table 5.4 and in Table 5.5. Generally, the oscillation can be reduced by decreasing the event triggering parameter ϵ_{CT}, which on the other hand results in an increased number of triggered events. An oscillation occurs also under the introduced event-triggering control methods due to the selection of $\epsilon \neq 0$ for both approaches. By selecting ϵ small enough the influence on the steady state performance is small, i.e. in this example the oscillation in the steady state leads to an output error $|y(t) - r| < 10^{-3}$ for both control input based and output based ETC resulting into a negligible increase of the cost function, see Table 5.5. On the other hand, the additional absolute threshold ϵ leads to an essential decrease of the triggered events. For instance under control input based ETC, 135 events are triggered for $\epsilon = 0$, see Table 5.1, whereas only 67 events

[1] For the continuous-time event-triggered PI control approach an input delay is not considered in the theoretical analysis. For comparison, the delay is included in the feedback link between the event-generator (EG) and the PI controller (T_{PI}) for simulating the communication time, see Figure 5.3. For the other approaches, the delay is included between the controller (T_{PI}) and the ZOH, assuming the controller is realized on the sensor side, see also the discussion at the end of Section 5.5 including Figure 5.17.

are triggered for $\epsilon = 3 \cdot 10^{-5}$ each with $\sigma = 0.1$. For both values of ϵ the costs are equal as indicated in Table 5.1 and Table 5.4. Under the simple ETC strategy a possible oscillation or sticking effect, respectively, is compensated by the safety measure of the event-triggering condition (5.71). However, this can result in an unnecessary utilization of the resources, see the number of triggered events in Table 5.5.

Example 5.2 Consider a second-order plant described by the continuous-time state equation

$$\begin{aligned} \dot{\boldsymbol{x}}_\mathrm{p}(t) &= \begin{pmatrix} -1 & 1 \\ 0 & -9 \end{pmatrix} \boldsymbol{x}_\mathrm{p}(t) + \begin{pmatrix} 0 \\ 4 \end{pmatrix} \boldsymbol{u}(t-\tau) \\ \boldsymbol{y}(t) &= \begin{pmatrix} 2.5 & 0 \end{pmatrix} \boldsymbol{x}_\mathrm{p}(t) \end{aligned} \tag{5.77}$$

with the initial state $\boldsymbol{x}_\mathrm{p}(0) = \boldsymbol{0}$. The plant describes the dynamic behavior of a satellite for the attitude control [DB05, Chapter 5]. The delay is assumed $\tau = 5\,\mathrm{ms}$ and the parameters of the continuous-time cost function are selected as $\boldsymbol{Q}_\mathrm{c} = \mathrm{diag}(20, 0)$, $Q_\mathrm{I} = 4.5$ and $R_\mathrm{c} = 0.3$. For applying the proposed event-triggered PI control strategy, the continuous-time state equation (5.76) and continuous-time cost function (5.7) are discretized with the step size $h = 10\,\mathrm{ms}$. In the following simulation a reference signal $r = 1$ is tracked and the simulation time is $T_\mathrm{eval} = 10\,\mathrm{s}$. For the control input based strategy the control parameters are designed by applying Theorem 5.2 with $\sigma = 0.3$. For the output based strategy the control synthesis (Theorem 5.4) is applied for $\sigma_0 = 0.01$ in order to get the control parameters. The event-triggered implementation is then realized with $\sigma = 0.2$, after verifying stability with Corollary 5.2. For both cases the parameter ϵ is selected as zero. An excerpt of the simulation results is shown in Figure 5.11 and the results are summarized in Table 5.6. In this example, the output based ETC approach leads to a better control performance due to the faster convergence to the setpoint,

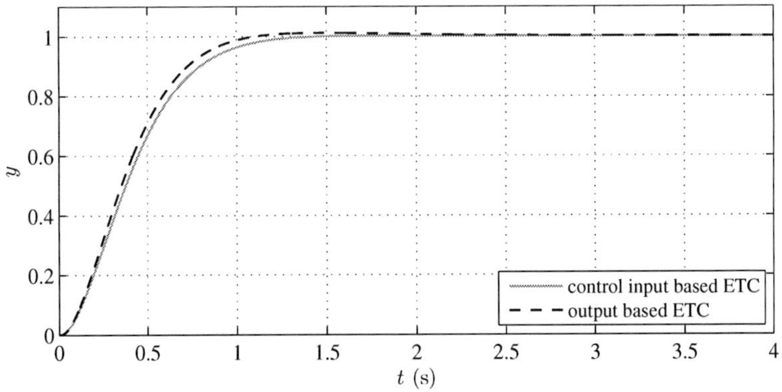

Figure 5.11: Output $y(t)$ under different event-triggered control strategies

Method	J_{eval}	N_{event}	$\boldsymbol{K}$
Control input based ETC	5.77	76	$(-1.9821 \quad -2.1003)$
Output based ETC	5.441	82	$(-2.1673 \quad -2.3513)$

Table 5.6: Evaluation under different event-triggered control strategies ($T_{\text{eval}} = 10\,\text{s}$)

whereas under the control input based ETC approach the resource utilization is lower. The example emphasizes the general applicability on different kind of systems.

The given analyses by simulation show that under the control input based event-triggering condition (5.11) the PI control synthesis allows designing efficiently the control parameters suitable for the event-triggered implementation of the PI controller. Under the output based event-triggering condition (5.12) the PI control synthesis shows good results for small values of σ. For increasing values of σ, conservatism appears, which can be compensated by selecting the control parameters according to proposed heuristic procedure. For both event-triggering conditions, the potential for significantly reducing the resource utilization while guaranteeing a certain control performance is clearly demonstrated.

5.5 Experimental Study

Besides the simulation examples considered in the previous section a practical verification of the proposed theory is crucial. The experimental study involves a DC-servo motor MAXON RE-max29 with an attached wheel as a plant and a microcontroller NXP LPC2294 as an implementation platform for the event generator as well as for the PI controller. The linear state equation of the plant is given by

$$\dot{n}(t) = -\frac{k_{\text{m}} k_{\text{e}}}{2\pi J R} n(t) + \frac{r_{\text{g}} k_{\text{m}}}{2\pi J R} u(t - \tau) \tag{5.78}$$

where $n(t)$ is the speed in rounds per second, $u(t)$ is the armature voltage in volt, and τ is the input delay that subsumes the required execution time of the sensoring, the event verification, the control computation and the actuation. The parameter $k_{\text{m}} = 0.0258\,\frac{\text{Nm}}{\text{A}}$ is the motor torque constant, $k_{\text{e}} = 0.1622\,\text{Vs}$ is the motor velocity constant, $R = 3.26\,\Omega$ is the resistance, $J = 8.79 \cdot 10^{-6}\,\text{kgm}^2$ is the moment of inertia, and $r_{\text{g}} = \frac{1}{28}$ is the gear ratio. The initial motor speed is $n(0) = 0\,\text{s}^{-1}$ and the control aim is to track a constant reference speed $r = 2\,\text{s}^{-1}$. With the state $x(t) = n(t)$, this leads to the plant model (5.1) with the corresponding parameters $\boldsymbol{A}_{\text{p}} = -23.2303$, $\boldsymbol{B}_{\text{p}} = 5.1162$ and $\boldsymbol{C}_{\text{p}} = 1$. Based on the given plant, the two proposed event-triggered PI control methods are experimentally evaluated in comparison with the simple event-triggered PI control approach and the continuous-time event-triggered PI control approach. The weighting matrices are selected as $Q_{\text{c}} = 1.5$, $Q_{\text{I}} = 25$ and $R_{\text{c}} = 0.1$ for designing the control parameters for all approaches and the evaluation time is chosen as $T_{\text{eval}} = 5\,\text{s}$. In the following the implementation of each method is discussed, while an overview of the results is given in Table 5.7. In the given plots an event is indicated by 1 and no event by 0.

Continuous-Time Event-Triggered PI Control

The continuous-time event-triggered PI control approach assumes a continuous monitoring of the output, which cannot be realized on the digital platform. Therefore, the event-triggering condition is verified periodically with a small step size $h = 1.6\,\text{ms}$ and the event-triggering condition (5.73) is adapted according to

$$\|\boldsymbol{y}(t) - \hat{\boldsymbol{y}}(t_k)\|_2 \geq \epsilon_{\text{CT}}. \tag{5.79}$$

The control gain $\boldsymbol{K} = \begin{pmatrix} -1.9246 & -15.8114 \end{pmatrix}$ is determined based on the continuous-time LQR method neglecting the input delay τ, see Appendix A.4.2. In Figure 5.12, the experimental results are given for two different values of the absolute threshold ϵ_{CT},

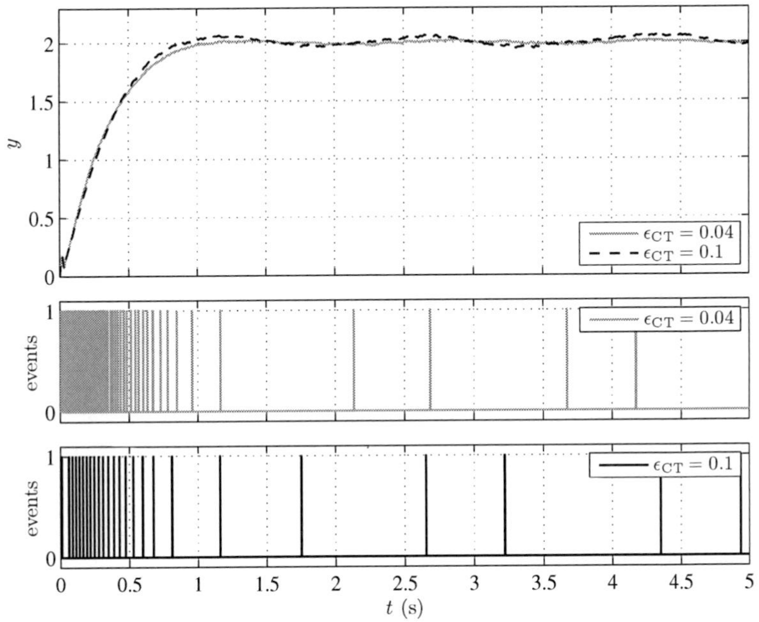

Figure 5.12: Output $y(t)$ under continuous-time event-triggered control

which clearly shows the influence of ϵ_{CT} on the number of events and the steady state performance. For $\epsilon_{\mathrm{CT}} = 0.04$ the output oscillates only slightly around the setpoint and 54 events are triggered. The number of events can be reduced to 26 for $\epsilon_{\mathrm{CT}} = 0.1$ at the cost of a severe oscillation of the output around the reference value. The resulting costs are similar for both cases, i.e. $J_{\mathrm{eval}} = 1.5147$ for $\epsilon_{\mathrm{CT}} = 0.04$ and $J_{\mathrm{eval}} = 1.5166$ for $\epsilon_{\mathrm{CT}} = 0.1$, as under the defined evaluation time the cost function J_{eval} values mainly the transient performance, and the oscillation has only small influence on the cost.

Simple Event-Triggered PI Control

For the remaining discrete-time methods the step size is chosen $h = 10\,\mathrm{ms}$. Based on Theorem A.2 the control gain is designed as $\boldsymbol{K} = \begin{pmatrix} -1.7782 & -15.0600 \end{pmatrix}$ for the simple event-triggered PI control strategy taking the input delay $\tau = 0.4662\,\mathrm{ms}$ into account.

Figure 5.13 shows the output behavior for $\epsilon_{\mathrm{DT}} = 0.03$ and $\epsilon_{\mathrm{DT}} = 0.1$ and $h_{\mathrm{max}} = 20h$ in both cases. For $\epsilon_{\mathrm{DT}} = 0.03$ the output converges faster to the reference whereas more events are triggered $N_{\mathrm{event}} = 68$. The resulting smoother control input leads to a smaller

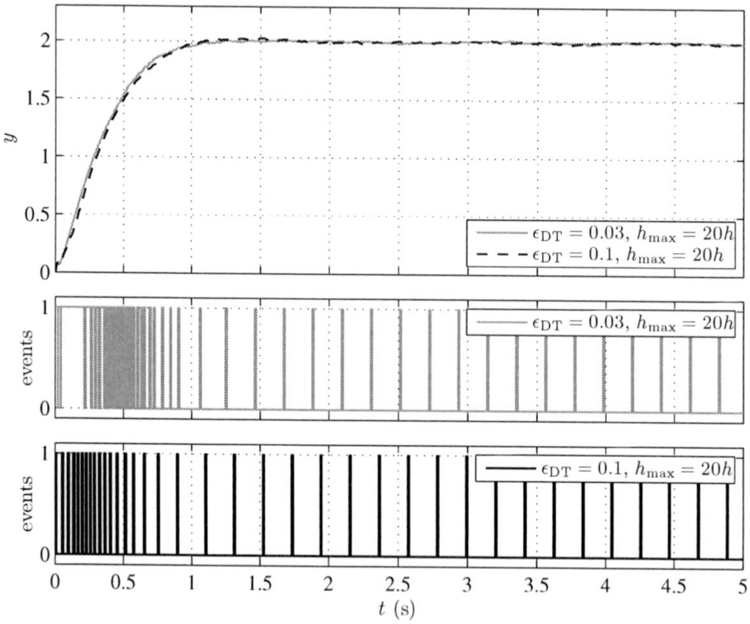

Figure 5.13: Output $y(t)$ under simple event-triggered control

cost $J_{\text{eval}} = 1.6614$ for $\epsilon_{\text{DT}} = 0.03$ than for $\epsilon_{\text{DT}} = 0.1$ where $J_{\text{eval}} = 1.9343$. For $\epsilon_{\text{DT}} = 0.1$ an event is triggered 37 times. During the steady state, events are triggered based on $h_{\max}$, which may lead to an unnecessary utilization of the resources.

Control Input Based Event-Triggered PI Control

The control synthesis (Theorem 5.2) is conducted with $\sigma = 0.5$, leading to the control gain $\boldsymbol{K} = \begin{pmatrix}-1.9127 & -17.3101\end{pmatrix}$. The measured execution time $\tau = 0.49781\,\text{ms}$ is considered in the control synthesis. Before the implementation the practical stability is proved based on Theorem 5.5 for $\epsilon = 0.001$. The event-triggered PI control approach is implemented for $\epsilon = 0$ and $\epsilon = 0.001$, leading to the results given in Figure 5.14. For both values of ϵ the control performance is improved compared with the previously given results, as the output converges faster to the reference value, see also Figure 5.16. This leads to smaller costs, which are $J_{\text{eval}} = 1.4949$ for $\epsilon = 0$ and $J_{\text{eval}} = 1.4896$ for $\epsilon = 0.001$. Additionally, the number of events is essentially reduced compared with the previous results, i.e. $N_{\text{event}} = 21$ for $\epsilon = 0$ and $N_{\text{event}} = 18$ for $\epsilon = 0.001$. The results

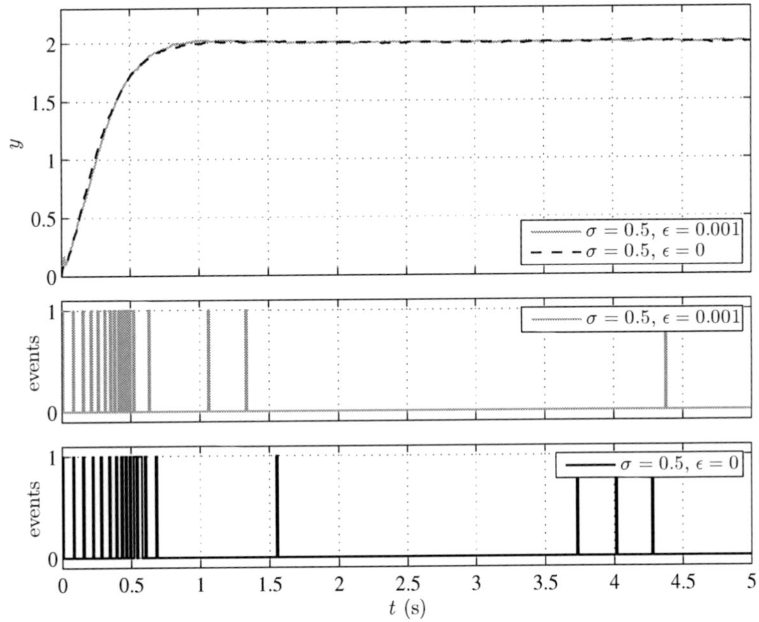

Figure 5.14: Output $y(t)$ under control input based event-triggered control

show the strength of the control synthesis, to design control parameters suitable for the event-triggered realization of a PI controller.

Output Based Event-Triggered PI Control

The control parameters are designed by using Theorem 5.4 for $\sigma_0 = 0.02$ with the measured execution time $\tau = 0.57141\,\text{ms}$, which results in $\boldsymbol{K} = \begin{pmatrix}-1.9111 & -15.0556\end{pmatrix}$. The implementation is then realized with $\sigma = 0.1$, which requires the stability verification based on Corollary 5.2. Furthermore, Theorem 5.6 is applied to analyze the stability for $\sigma = 0.1$ and $\epsilon = 0.0002$ with the designed control gain. Figure 5.15 shows especially how the additionally absolute threshold ϵ can reduce the number of events while maintaining the control performance, which is $J_{\text{eval}} = 1.6562$ for $\epsilon = 0$ and $J_{\text{eval}} = 1.6553$ for $\epsilon = 0.0002$. For $\epsilon = 0$ there are 322 events, while $\epsilon = 0.0002$ leads to only 44 events.

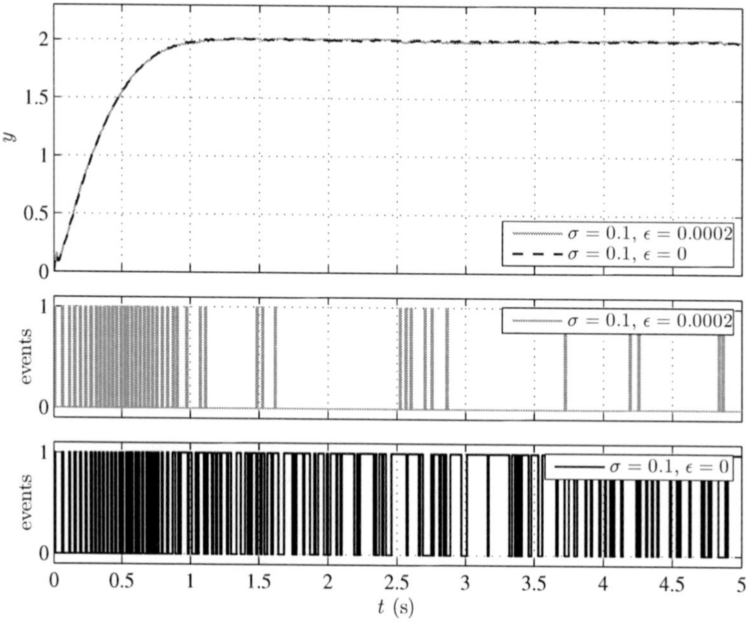

Figure 5.15: Output $y(t)$ under output based event-triggered control

Comparison and Discussion

Figure 5.16 shows the experimental results of all approaches. As already highlighted the control input based ETC strategy converges faster to the reference value, which also leads to the smallest cost J_{eval}, see Table 5.7. The other three strategies show a very similar output performance, whereas the continuous-time approach leads to a smaller cost J_{eval}. This can be explained by the smooth control input, which is updated continuously both when an event is triggered and otherwise under the continuous-time approach. For the simple ETC and the output based ETC approach, the control input is step-like resulting in a higher cost. Essential differences among the methods can be found in the number of triggered events. The number of triggered events for the proposed control input based strategy is the smallest, while realizing the best control performance. For the output

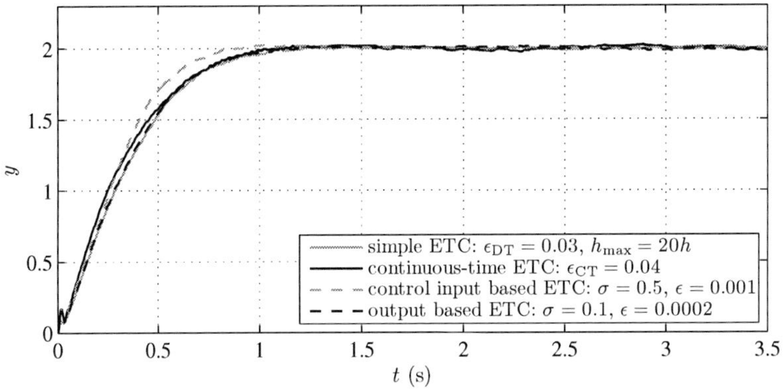

Figure 5.16: Excerpt of the output $y(t)$ under different ETC strategies

Method	J_{eval}	N_{event}
Continuous-time ETC: $\epsilon_{\text{CT}=0.04}$	1.5147	54
Continuous-time ETC: $\epsilon_{\text{CT}=0.1}$	1.5166	26
Simple ETC: $\epsilon_{\text{DT}} = 0.03$, $h_{\max} = 20h$	1.6614	68
Simple ETC: $\epsilon_{\text{DT}} = 0.1$, $h_{\max} = 20h$	1.9343	37
Control input based ETC: $\sigma = 0.5$, $\epsilon = 0$	1.4949	21
Control input based ETC: $\sigma = 0.5$, $\epsilon = 0.001$	1.4896	18
Output based ETC: $\sigma = 0.1$, $\epsilon = 0$	1.6562	322
Output based ETC: $\sigma = 0.1$, $\epsilon = 0.0002$	1.6553	44

Table 5.7: Experimental evaluation under different ETC strategies ($T_{\text{eval}} = 5\,\text{s}$)

based ETC approach the number of triggered events can be kept comparatively small, due to the usage of the absolute threshold ϵ, while maintaining the control performance. For the continuous-time ETC and the simple ETC approach a smaller number of events can only be achieved at the cost of an essential degradation of the control performance in terms of a severe oscillation under continuous-time ETC (Figure 5.12) and in terms of an increase of the cost J_{eval} under simple ETC (Table 5.7).

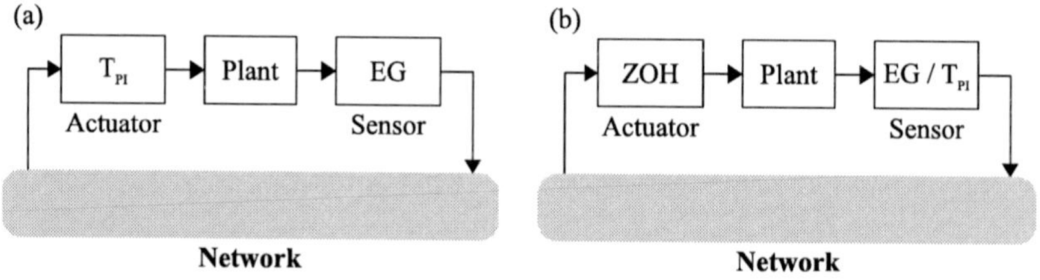

Figure 5.17: Topologies for networked event-triggered control systems

In the context of an application of the discussed methods also other aspects may be relevant, which are related to the structure of the methods. The simple event-triggered control approach benefits from its simplicity and the low computational complexity of the event-triggering condition. Therefore, it is also applicable for saving computation resources. Another advantage is that in a networked control system the PI controller (T_{PI}) can be realized both on the sensor side and the actuator side, see Figure 5.17. For the continuous-time approach, the PI controller needs to be implemented on the actuator side, see Figure 5.17 (a), as the control input is updated continuously. Further, actuator saturation and anti-windup compensation are already analyzed for the continuous-time approach [LKJ12], which remains an open point for the discrete-time methods. Compared with the other methods, the proposed event-triggered control methods require the computation of the PI controller or the integration of the output error for verifying the event-triggering condition. Thus, it is crucial to shift the computation of the PI controller to the sensor side, see Figure 5.17 (b).

In summary, also the experimental study can show the efficiency of the PI control synthesis under the control input based event-triggering condition (5.11). The output based event-triggering condition (5.12) demonstrates the benefits of an event-triggering condition combining a relative and an absolute threshold. Regarding both the control performance and resources savings the introduced approaches are promising in comparison with existing approaches, especially in practical implementations.

5.6 Summary

In this chapter, a novel event-triggered PI control synthesis is proposed, which allows to design the control parameters suitable for the event-triggered implementation. Based on the developed PI control structure, challenges as oscillation and sticking can be handled efficiently. Further, methods for analyzing stability are introduced. The effectiveness of the presented strategy is evaluated by simulations and experimental studies in comparison with existing strategies in the literature. The proposed event-triggered PI control strategy shows its ability of utilizing the resources more efficiently while providing similar control performance.

6 Conclusions and Outlook

6.1 Conclusions

In this thesis, several strategies have been studied for resource-constrained control systems. In such systems only limited computation and communication resources are available for control. This challenge has been tackled by two strategies. In Part I two scheduling strategies have been presented for distributing efficiently the resources. The idea of event-triggered control has been utilized in Part II to reduce the utilization of communication resources while maintaining a certain control performance.

One scheduling approach to distributing limited communication resources has been presented in Chapter 3. A joint control and scheduling design has been developed considering a linear quadratic performance measure. The control and scheduling based on output-feedback information has been realized by including a switched observer in the scheme. Evaluations have been made based on simulation and experimental studies, which have clearly shown the efficiency of the proposed strategy.

Another strategy for the scheduling of a set of PI control tasks with limited computation resources has been introduced in Chapter 4. Besides a novel online scheduling approach based on output-feedback information, a strategy for designing the control parameters has been presented. The benefits of the proposed method have been demonstrated by comparing it with the EDF scheduling and an offline scheduling approach. The comparisons have been done through simulative and experimental studies.

In Chapter 5, a novel event-triggered PI control concept has been proposed for setpoint tracking problems. The strategy follows the concept to verify the event-triggering condition only periodically and not continuously, such that an implementation can be easily realized in a standard time-sliced embedded software architecture. The event-triggering condition is designed in a way, such that challenges of event-triggered PI control such as oscillation and sticking can be handled. Finally, a control synthesis has been presented which allows to design the control parameters taking the event-triggered implementation into account. The simulations and practical implementation have shown that the communication effort can be strongly reduced, while keeping a similar control performance. Also in comparison with other approaches from literature, the benefits of the proposed strategy have been shown.

6.2 Outlook

Several concepts for dealing with resource-constrained control systems have been discussed in this thesis. The ideas can be extended to the following research directions.

Extended scheduling of the communication network

For the online scheduling strategy in Chapter 3, it is assumed the state or output information of all plants is available for executing the scheduler. If this information needs to be transmitted over the network, this leads to a considerable communication overhead. Motivated by this, a heuristic strategy taking the scheduling overhead into account is presented in [RAL13a], where scheduler switches between offline and online scheduling based on certain conditions. However, the quantitative assessment of the achievable control performance remains an open problem.

Alternatively, the scheduling can be realized in a decentralized manner. In [AGRL13, AGL15] for instance, a decentralized event-triggered scheduler is presented. The proposed scheduling can be implemented in a CAN bus protocol with dynamic priorities, where the priorities are determined by an event-triggering law for each sensor node. Besides an efficient distribution of the limited resources, this research direction combining scheduling strategies and event-triggered control allows also to reduce the resource utilization. Further results on event-triggered control of multi-loop systems are presented in [CH08], where the control performance is analyzed for various collision-free medium access protocols, or in [MH11, MH14], where the joint design of the event-triggered controller and scheduler is realized in the framework of stochastic optimal control. An interesting open point is the analysis of linear dynamic controllers in the context of event-triggered control and scheduling.

Scheduling of control tasks of high computational complexity

The online scheduling of control tasks for controlling a set of plants has been discussed in Chapter 4 for the case, where the control task consists of a PI controller. An extension to control tasks of higher computational complexity may be interesting, for instant model predictive control tasks. Thereby, the challenge involves finding a scheduling law of low complexity, which allows to distribute the computation resources online in an efficient manner.

Event-triggered PI control with actuator saturation

The event-triggered PI control strategy, introduced in Chapter 5, can handle already several challenges, which appear in the event-triggered realization of PI controllers. However, the possible saturation of the control input is not considered, which is a common problem in practical systems. For discrete-time event-triggered control systems which are controlled by a state-feedback control law such problems are investigated in [SPTZ13, WRL14]. In [LKJ12], actuator saturation and anti-windup compensation are analyzed for event-triggered PI control, where the event-triggering condition is monitored continuously. An extension of this approach leading to a periodic monitoring of the event-triggering condition is of interest from the point of practical implementation.

Extension towards communication imperfections

The discussed approaches in Chapter 3 and 5 consider the efficient utilization of the limited communication resources. For both approaches, a constant communication delay is included in the modeling. However, the transmission times are often uncertain, which leads to a time-varying delay. Additionally, packet loss and quantization effects may occur. In literature, such communication imperfections have been widely studied in the context of networked control systems, see e.g. [HNX07, NFZE07, DHVH11]. Based on existing methods, future work may therefore focus on extending the proposed strategies, such that the communication imperfections are explicitly considered in the design procedure. In this context, the practical implementation of the methods in a setup where the communication is realized over a network is also of interest.

Part III

Appendix

A Supplementary Material

A.1 Discretization of the Cost Function

This appendix addresses the discretization of the continuous-time cost functions (3.2), (4.20) and (5.7). For more details on the discretization of the cost function, see [ÅW90, Section 11.1], [San04, Paper B, Section 5.3] and [Gör12, Appendix A.4].

A.1.1 For Control and Scheduling Codesign (Chapter 3)

The continuous-time cost function (3.2) is discretized over the discretization interval $t_k \leq t < t_{k+1}$ using zero-order hold. Thereby, it is distinguished if the control task T_i of the considered plant P_i is executed or not within $t_k \leq t < t_{k+1}$. The cost function (3.2) can equivalently be written as

$$J_i = \sum_{k=1}^{\infty} \int_{t_k}^{t_{k+1}} \begin{pmatrix} \boldsymbol{x}_{\mathrm{p}i}(t) \\ \boldsymbol{u}_i(t-\tau_i) \end{pmatrix}^T \begin{pmatrix} \boldsymbol{Q}_{\mathrm{c}i} & \mathbf{0} \\ \mathbf{0} & \boldsymbol{R}_{\mathrm{c}i} \end{pmatrix} \begin{pmatrix} \boldsymbol{x}_{\mathrm{p}i}(t) \\ \boldsymbol{u}_i(t-\tau_i) \end{pmatrix} dt \tag{A.1}$$

with $t_{k+1} - t_k = h_{j(k)}$. Substituting the solution of the continuous-time state equation (3.1) into (A.1) and calculating the integrals leads to the discrete-time cost function

$$J_i = \sum_{k=0}^{\infty} \begin{pmatrix} \boldsymbol{x}_{\mathrm{p}i}(k) \\ \boldsymbol{u}_i(k-1) \\ \boldsymbol{u}_i(k) \end{pmatrix}^T \underbrace{\begin{pmatrix} \boldsymbol{Q}_{11ij(k)} & \boldsymbol{Q}_{12ij(k)} & \boldsymbol{Q}_{13ij(k)} \\ * & \boldsymbol{Q}_{22ij(k)} & \boldsymbol{Q}_{23ij(k)} \\ * & * & \boldsymbol{Q}_{33ij(k)} \end{pmatrix}}_{\boldsymbol{Q}_{ij(k)}} \begin{pmatrix} \boldsymbol{x}_{\mathrm{p}i}(k) \\ \boldsymbol{u}_i(k-1) \\ \boldsymbol{u}_i(k) \end{pmatrix} \tag{A.2}$$

associated to the plant P_i with

$$\boldsymbol{Q}_{11ij(k)} = \int_0^{h_{j(k)}} \boldsymbol{\Phi}_i^T(t)\boldsymbol{Q}_{\mathrm{c}i}\boldsymbol{\Phi}_i(t)dt \tag{A.3a}$$

$$\boldsymbol{Q}_{12ij(k)} = \int_0^{\check{\tau}_{ij(k)}} \boldsymbol{\Phi}_i^T(t)\boldsymbol{Q}_{\mathrm{c}i}\boldsymbol{\Gamma}_i(0,t)dt + \int_{\check{\tau}_{ij(k)}}^{h_{j(k)}} \boldsymbol{\Phi}_i^T(t)\boldsymbol{Q}_{\mathrm{c}i}\boldsymbol{\Gamma}_i(0,\check{\tau}_{ij(k)})dt \tag{A.3b}$$

$$\boldsymbol{Q}_{13ij(k)} = \int_{\check{\tau}_{ij(k)}}^{h_{j(k)}} \boldsymbol{\Phi}_i^T(t)\boldsymbol{Q}_{\mathrm{c}i}\boldsymbol{\Gamma}_i(\check{\tau}_{ij(k)},t)dt \tag{A.3c}$$

$$\boldsymbol{Q}_{22ij(k)} = \int_0^{\check{\tau}_{ij(k)}} \boldsymbol{\Gamma}_i^T(0,t)\boldsymbol{Q}_{ci}\boldsymbol{\Gamma}_i(0,t) + \boldsymbol{R}_{ci}dt + \int_{\check{\tau}_{ij(k)}}^{h_{j(k)}} \boldsymbol{\Gamma}_i^T(0,\check{\tau}_{ij(k)})\boldsymbol{Q}_{ci}\boldsymbol{\Gamma}_i(0,\check{\tau}_{ij(k)})dt \quad \text{(A.3d)}$$

$$\boldsymbol{Q}_{23ij(k)} = \int_{\check{\tau}_{ij(k)}}^{h_{j(k)}} \boldsymbol{\Gamma}_i^T(0,\check{\tau}_{ij(k)})\boldsymbol{Q}_{ci}\boldsymbol{\Gamma}_i(\check{\tau}_{ij(k)},t)dt \quad \text{(A.3e)}$$

$$\boldsymbol{Q}_{33ij(k)} = \int_{\check{\tau}_{ij(k)}}^{h_{j(k)}} \boldsymbol{\Gamma}_i^T(\check{\tau}_{ij(k)},t)\boldsymbol{Q}_{ci}\boldsymbol{\Gamma}_i(\check{\tau}_{ij(k)},t) + \boldsymbol{R}_{ci}dt \quad \text{(A.3f)}$$

where

$$\boldsymbol{\Phi}_i(t) = e^{\boldsymbol{A}_{\mathrm{p}i}t}, \qquad \boldsymbol{\Gamma}_i(t_1,t_2) = \int_{t_1}^{t_2} e^{\boldsymbol{A}_{\mathrm{p}i}(t-s)}\boldsymbol{B}_{\mathrm{p}i}ds.$$

Due to the required distinction if the control task T_i of the considered plant P_i is executed or not, the weighting matrix $\boldsymbol{Q}_{ij(k)}$ contains the switching index $j(k)$. The overall cost function, the sum of the individual cost functions, is given by

$$J = \sum_{i=1}^{M} J_i = \sum_{k=0}^{\infty} \begin{pmatrix} \boldsymbol{x}(k) \\ \boldsymbol{u}(k) \end{pmatrix}^T \underbrace{\begin{pmatrix} \tilde{\boldsymbol{Q}}_{1j(k)} & \tilde{\boldsymbol{Q}}_{12j(k)} \\ * & \tilde{\boldsymbol{Q}}_{2j(k)} \end{pmatrix}}_{\boldsymbol{Q}_{j(k)}} \begin{pmatrix} \boldsymbol{x}(k) \\ \boldsymbol{u}(k) \end{pmatrix} \quad \text{(A.4)}$$

where the block-diagonal weighting matrices are given by

$$\tilde{\boldsymbol{Q}}_{1j(k)} = \operatorname{diag}\left(\begin{pmatrix} \boldsymbol{Q}_{111j(k)} & \boldsymbol{Q}_{121j(k)} \\ * & \boldsymbol{Q}_{221j(k)} \end{pmatrix}, \dots, \begin{pmatrix} \boldsymbol{Q}_{11Mj(k)} & \boldsymbol{Q}_{12Mj(k)} \\ * & \boldsymbol{Q}_{22Mj(k)} \end{pmatrix}\right) \quad \text{(A.5a)}$$

$$\tilde{\boldsymbol{Q}}_{12j(k)} = \operatorname{diag}\left(\begin{pmatrix} \boldsymbol{Q}_{131j(k)} \\ \boldsymbol{Q}_{231j(k)} \end{pmatrix}, \dots, \begin{pmatrix} \boldsymbol{Q}_{13Mj(k)} \\ \boldsymbol{Q}_{23Mj(k)} \end{pmatrix}\right) \quad \text{(A.5b)}$$

$$\tilde{\boldsymbol{Q}}_{2j(k)} = \operatorname{diag}\left(\boldsymbol{Q}_{331j(k)}, \dots, \boldsymbol{Q}_{33Mj(k)}\right). \quad \text{(A.5c)}$$

A.1.2 For PI Control and Scheduling (Chapter 4)

For discretizing the continuous-time cost function (4.20) over the discretization interval $t_{k_i} \le t < t_{k_i+1}$ using zero-order hold, the cost function is rewritten as

$$J_i = \sum_{k_i=1}^{\infty} \int_{t_{k_i}}^{t_{k_i+1}} \left(\begin{pmatrix} \boldsymbol{x}_{\mathrm{p}i}(t) \\ \boldsymbol{u}_i(t-\tau) \end{pmatrix}^T \begin{pmatrix} \boldsymbol{Q}_{ci} & \boldsymbol{0} \\ \boldsymbol{0} & \boldsymbol{R}_{ci} \end{pmatrix} \begin{pmatrix} \boldsymbol{x}_{\mathrm{p}i}(t) \\ \boldsymbol{u}_i(t-\tau) \end{pmatrix} + \boldsymbol{x}_{\mathrm{I}i}^T(t)\boldsymbol{Q}_{\mathrm{I}i}\boldsymbol{x}_{\mathrm{I}i}(t) \right) dt. \quad \text{(A.6)}$$

As the PI controller is realized on a digital platform, the integrator state is not updated continuously, i.e. $\boldsymbol{x}_{\mathrm{I}i}(t) = \boldsymbol{x}_{\mathrm{I}i}(t_{k_i})$ for $t \in [t_{k_i}, t_{k_i+1})$. As the reference signal is assumed zero, the integration of the integrator state is given by $\boldsymbol{x}_{\mathrm{I}i}(k_i) = \boldsymbol{x}_{\mathrm{I}i}(k_i-1) + h\boldsymbol{y}_i(k_i)$. Thus, (A.6) can be rewritten as

$$\begin{aligned} J_i = \sum_{k_i=1}^{\infty} \Bigg(\int_{t_{k_i}}^{t_{k_i+1}} \begin{pmatrix} \boldsymbol{x}_{\mathrm{p}i}(t) \\ \boldsymbol{u}_i(t-\tau) \end{pmatrix}^T \begin{pmatrix} \boldsymbol{Q}_{ci} & \boldsymbol{0} \\ \boldsymbol{0} & \boldsymbol{R}_{ci} \end{pmatrix} \begin{pmatrix} \boldsymbol{x}_{\mathrm{p}i}(t) \\ \boldsymbol{u}_i(t-\tau) \end{pmatrix} dt \Bigg) \\ + \left(\boldsymbol{x}_{\mathrm{I}i}(k_i-1) + h\boldsymbol{y}_i(k_i)\right)^T \boldsymbol{Q}_{\mathrm{I}i} \left(\boldsymbol{x}_{\mathrm{I}i}(k_i-1) + h\boldsymbol{y}_i(k_i)\right) \end{aligned} \quad \text{(A.7)}$$

which is equivalent to

$$
J_i = \sum_{k_i=0}^{\infty} \left(\int_{t_{k_i}}^{t_{k_i+1}} \begin{pmatrix} \boldsymbol{x}_{\mathrm{p}i}(t) \\ \boldsymbol{u}_i(t-\tau) \end{pmatrix}^T \begin{pmatrix} \boldsymbol{Q}_{ci} & \boldsymbol{0} \\ \boldsymbol{0} & \boldsymbol{R}_{ci} \end{pmatrix} \begin{pmatrix} \boldsymbol{x}_{\mathrm{p}i}(t) \\ \boldsymbol{u}_i(t-\tau) \end{pmatrix} dt \right)
$$
$$
+ \begin{pmatrix} \boldsymbol{x}_{\mathrm{p}i}(k_i) \\ \boldsymbol{x}_{\mathrm{I}i}(k_i-1) \end{pmatrix}^T \begin{pmatrix} h^2 \boldsymbol{C}_{\mathrm{p}i}^T \boldsymbol{Q}_{\mathrm{I}i} \boldsymbol{C}_{\mathrm{p}i} & h \boldsymbol{C}_{\mathrm{p}i}^T \boldsymbol{Q}_{\mathrm{I}i} \\ * & \boldsymbol{Q}_{\mathrm{I}i} \end{pmatrix} \begin{pmatrix} \boldsymbol{x}_{\mathrm{p}i}(k_i) \\ \boldsymbol{x}_{\mathrm{I}i}(k_i-1) \end{pmatrix} \tag{A.8}
$$

where $h = t_{k_i+1} - t_{k_i}$. This leads to the discretized cost function (4.21)

$$
J_i = \sum_{k_i=0}^{\infty} \begin{pmatrix} \boldsymbol{x}_{\mathrm{p}i}(k_i) \\ \boldsymbol{u}_i(k_i-1) \\ \boldsymbol{x}_{\mathrm{I}i}(k_i-1) \\ \boldsymbol{u}_i(k_i) \end{pmatrix}^T \boldsymbol{Q}_i \begin{pmatrix} \boldsymbol{x}_{\mathrm{p}i}(k_i) \\ \boldsymbol{u}_i(k_i-1) \\ \boldsymbol{x}_{\mathrm{I}i}(k_i-1) \\ \boldsymbol{u}_i(k_i) \end{pmatrix} \tag{A.9}
$$

with

$$
\boldsymbol{Q}_i = \begin{pmatrix} \boldsymbol{Q}_{11i} + \boldsymbol{Q}_{\mathrm{I}11i} & \boldsymbol{Q}_{12i} & \boldsymbol{Q}_{\mathrm{I}13i} & \boldsymbol{Q}_{14i} \\ * & \boldsymbol{Q}_{22i} & \boldsymbol{0} & \boldsymbol{Q}_{24i} \\ * & * & \boldsymbol{Q}_{\mathrm{I}33i} & \boldsymbol{0} \\ * & * & * & \boldsymbol{Q}_{44i} \end{pmatrix} \tag{A.10}
$$

where

$$
\boldsymbol{Q}_{11i} = \int_0^h \boldsymbol{\Phi}_i^T(t) \boldsymbol{Q}_{ci} \boldsymbol{\Phi}_i(t) dt \tag{A.11a}
$$

$$
\boldsymbol{Q}_{12i} = \int_0^{\tau_i} \boldsymbol{\Phi}_i^T(t) \boldsymbol{Q}_{ci} \boldsymbol{\Gamma}_i(0,t) dt + \int_{\tau_i}^h \boldsymbol{\Phi}_i^T(t) \boldsymbol{Q}_{ci} \boldsymbol{\Gamma}_i(0,\tau_i) dt \tag{A.11b}
$$

$$
\boldsymbol{Q}_{14i} = \int_{\tilde{\tau}_i}^h \boldsymbol{\Phi}_i^T(t) \boldsymbol{Q}_{ci} \boldsymbol{\Gamma}_i(\tau_i,t) dt \tag{A.11c}
$$

$$
\boldsymbol{Q}_{22i} = \int_0^{\tau_i} \boldsymbol{\Gamma}_i^T(0,t) \boldsymbol{Q}_{ci} \boldsymbol{\Gamma}_i(0,t) + \boldsymbol{R}_{ci} dt + \int_{\tau_i}^h \boldsymbol{\Gamma}_i^T(0,\tau_i) \boldsymbol{Q}_{ci} \boldsymbol{\Gamma}_i(0,\tau_i) dt \tag{A.11d}
$$

$$
\boldsymbol{Q}_{24i} = \int_{\tau_i}^h \boldsymbol{\Gamma}_i^T(0,\tau_i) \boldsymbol{Q}_{ci} \boldsymbol{\Gamma}_i(\tau_i,t) dt \tag{A.11e}
$$

$$
\boldsymbol{Q}_{44i} = \int_{\tau_i}^h \boldsymbol{\Gamma}_i^T(\tau_i,t) \boldsymbol{Q}_{ci} \boldsymbol{\Gamma}_i(\tau_i,t) + \boldsymbol{R}_{ci} dt \tag{A.11f}
$$

$$
\boldsymbol{Q}_{\mathrm{I}11i} = h^3 \cdot \boldsymbol{C}_{\mathrm{p}i}^T \boldsymbol{Q}_{\mathrm{I}i} \boldsymbol{C}_{\mathrm{p}i} \tag{A.11g}
$$

$$
\boldsymbol{Q}_{\mathrm{I}13i} = h^2 \cdot \boldsymbol{C}_{\mathrm{p}i}^T \boldsymbol{Q}_{\mathrm{I}i} \tag{A.11h}
$$

$$
\boldsymbol{Q}_{\mathrm{I}33i} = h \cdot \boldsymbol{Q}_{\mathrm{I}i} \tag{A.11i}
$$

and

$$
\boldsymbol{\Phi}_i(t) = e^{\boldsymbol{A}_{\mathrm{p}i} t}, \qquad \boldsymbol{\Gamma}_i(t_1,t_2) = \int_{t_1}^{t_2} e^{\boldsymbol{A}_{\mathrm{p}i}(t-s)} \boldsymbol{B}_{\mathrm{p}i} ds.
$$

A.1.3 For Event-Triggered PI Control (Chapter 5)

For the discretization of the continuous-time cost function (5.7) over the discretization interval $t_k \leq t < t_{k+1}$ using zero-order hold, the cost function is rewritten as

$$J = \sum_{k=1}^{\infty} \int_{t_k}^{t_{k+1}} \left(\begin{pmatrix} \tilde{\boldsymbol{x}}_{\mathrm{p}}(t) \\ \tilde{\boldsymbol{u}}(t-\tau) \end{pmatrix}^T \begin{pmatrix} \boldsymbol{Q}_{\mathrm{c}} & 0 \\ 0 & \boldsymbol{R}_{\mathrm{c}} \end{pmatrix} \begin{pmatrix} \tilde{\boldsymbol{x}}_{\mathrm{p}}(t) \\ \tilde{\boldsymbol{u}}(t-\tau) \end{pmatrix} + \tilde{\boldsymbol{x}}_{\mathrm{I}}^T(t) \boldsymbol{Q}_{\mathrm{I}} \tilde{\boldsymbol{x}}_{\mathrm{I}}(t) \right) dt. \quad \text{(A.12)}$$

As the PI controller is realized on a digital platform, the integrator state is not updated continuously, i.e. $\tilde{\boldsymbol{x}}_{\mathrm{I}}(t) = \tilde{\boldsymbol{x}}_{\mathrm{I}}(t_k)$ for $t \in [t_k, t_{k+1})$. Therefore, (A.12) is equivalent to

$$J = \sum_{k=1}^{\infty} \left(\int_{t_k}^{t_{k+1}} \begin{pmatrix} \tilde{\boldsymbol{x}}_{\mathrm{p}}(t) \\ \tilde{\boldsymbol{u}}(t-\tau) \end{pmatrix}^T \begin{pmatrix} \boldsymbol{Q}_{\mathrm{c}} & 0 \\ 0 & \boldsymbol{R}_{\mathrm{c}} \end{pmatrix} \begin{pmatrix} \tilde{\boldsymbol{x}}_{\mathrm{p}}(t) \\ \tilde{\boldsymbol{u}}(t-\tau) \end{pmatrix} dt + \tilde{\boldsymbol{x}}_{\mathrm{I}}^T(t_k)(h\boldsymbol{Q}_{\mathrm{I}})\tilde{\boldsymbol{x}}_{\mathrm{I}}(t_k) \right) \quad \text{(A.13)}$$

where $h = t_{k+1} - t_k$. This leads to the discretized cost function (5.10)

$$J = \sum_{k=0}^{\infty} \begin{pmatrix} \tilde{\boldsymbol{x}}_{\mathrm{p}}(k) \\ \tilde{\boldsymbol{u}}(k-1) \\ \tilde{\boldsymbol{x}}_{\mathrm{I}}(k) \\ \tilde{\boldsymbol{u}}(k) \end{pmatrix}^T \boldsymbol{Q} \begin{pmatrix} \tilde{\boldsymbol{x}}_{\mathrm{p}}(k) \\ \tilde{\boldsymbol{u}}(k-1) \\ \tilde{\boldsymbol{x}}_{\mathrm{I}}(k) \\ \tilde{\boldsymbol{u}}(k) \end{pmatrix} \quad \text{(A.14)}$$

with

$$\boldsymbol{Q} = \begin{pmatrix} \boldsymbol{Q}_{11} & \boldsymbol{Q}_{12} & 0 & \boldsymbol{Q}_{14} \\ * & \boldsymbol{Q}_{22} & 0 & \boldsymbol{Q}_{24} \\ * & * & \boldsymbol{Q}_{33} & 0 \\ * & * & * & \boldsymbol{Q}_{44} \end{pmatrix} \quad \text{(A.15)}$$

where

$$\boldsymbol{Q}_{11} = \int_0^h \boldsymbol{\Phi}^T(t)\boldsymbol{Q}_{\mathrm{c}}\boldsymbol{\Phi}(t)dt \quad \text{(A.16a)}$$

$$\boldsymbol{Q}_{12} = \int_0^\tau \boldsymbol{\Phi}^T(t)\boldsymbol{Q}_{\mathrm{c}}\boldsymbol{\Gamma}(0,t)dt + \int_\tau^h \boldsymbol{\Phi}^T(t)\boldsymbol{Q}_{\mathrm{c}}\boldsymbol{\Gamma}(0,\tau)dt \quad \text{(A.16b)}$$

$$\boldsymbol{Q}_{14} = \int_\tau^h \boldsymbol{\Phi}^T(t)\boldsymbol{Q}_{\mathrm{c}}\boldsymbol{\Gamma}(\tau,t)dt \quad \text{(A.16c)}$$

$$\boldsymbol{Q}_{22} = \int_0^\tau \boldsymbol{\Gamma}^T(0,t)\boldsymbol{Q}_{\mathrm{c}}\boldsymbol{\Gamma}(0,t) + \boldsymbol{R}_{\mathrm{c}}dt + \int_\tau^h \boldsymbol{\Gamma}^T(0,\tau)\boldsymbol{Q}_{\mathrm{c}}\boldsymbol{\Gamma}(0,\tau)dt \quad \text{(A.16d)}$$

$$\boldsymbol{Q}_{24} = \int_\tau^h \boldsymbol{\Gamma}^T(0,\tau)\boldsymbol{Q}_{\mathrm{c}}\boldsymbol{\Gamma}(\tau,t)dt \quad \text{(A.16e)}$$

$$\boldsymbol{Q}_{44} = \int_\tau^h \boldsymbol{\Gamma}^T(\tau,t)\boldsymbol{Q}_{\mathrm{c}}\boldsymbol{\Gamma}(\tau,t) + \boldsymbol{R}_{\mathrm{c}}dt \quad \text{(A.16f)}$$

$$\boldsymbol{Q}_{33} = h\boldsymbol{Q}_{\mathrm{I}} \quad \text{(A.16g)}$$

and

$$\boldsymbol{\Phi}(t) = e^{\boldsymbol{A}_{\mathrm{p}}t}, \qquad \boldsymbol{\Gamma}(t_1,t_2) = \int_{t_1}^{t_2} e^{\boldsymbol{A}_{\mathrm{p}}(t-s)}\boldsymbol{B}_{\mathrm{p}}ds.$$

A.2 Measurement of the Execution Times of the Experiment in Section 3.7

Table A.1 and Table A.2 give the measured execution times for the experiments presented in Section 3.7. The execution times refer to the different steps of realization discussed in Chapter 2.

Method	Online scheduling		Offline scheduling	
Control Task	T_1, T_2	T_3	T_1, T_2	T_3
Input reading	850.3	850.3	208.9	432.5
Scheduling	1467.4	1467.4	0	0
Control computation	47.0	58.5	47.0	58.5
Output application	129.9	129.9	129.9	129.9
Total	2494.6	2506.1	385.5	620.9

Table A.1: Measured execution times of the segments of the control task T_i in microseconds for one inverted pendulum and two DC motors

Method	SF-on	OF-PO-on	OF-CO-on
Input reading	167.6	127.4	127.4
Observer (correction)	0	0	113.9
Scheduling	539.8	539.8	539.8
Control computation	101.8	101.8	101.8
Output application	32.1	32.1	32.1
Observer (prediction)	0	298.0	206.7
Total	841.3	1099.1	1121.7

Table A.2: Measured execution times of the segments of the control task T_i in microseconds for two double integrator circuits

A.3 Control Design for the Related Methods of Scheduling of PI Control Tasks

For comparison, the scheduling of PI control tasks is realized under non-preemptive offline scheduling and under preemptive EDF scheduling. For both cases the tasks are executed periodically with a constant period h_{act_i}. The control parameters are equivalently designed minimizing an upper bound of the cost function (4.20). The PI controller is realized with forward integration, see equation (4.52), which leads to the closed-loop model of a plant P_i

$$\boldsymbol{z}_i(k_i+1) = \left(\boldsymbol{A}_{\mathrm{z}i}(h_{\mathrm{act}_i}) + \boldsymbol{B}_{\mathrm{z}i}(h_{\mathrm{act}_i})\boldsymbol{K}_i\boldsymbol{C}_{\mathrm{z}i}(h_{\mathrm{act}_i})\right)\boldsymbol{z}_i(k_i), \tag{A.17}$$

with $\boldsymbol{z}_i(k_i) = \left(\boldsymbol{x}_{\mathrm{a}i}^T(k_i) \quad \boldsymbol{x}_{\mathrm{I}i}^T(k_i)\right)^T$, $\boldsymbol{K}_i = \left(\boldsymbol{K}_{\mathrm{P}i} \quad \boldsymbol{K}_{\mathrm{I}i}\right)$ and the matrices

$$\boldsymbol{A}_{\mathrm{z}i}(h_{\mathrm{act}_i}) = \begin{pmatrix} \boldsymbol{A}_{\mathrm{a}i}(h_{\mathrm{act}_i}) & 0 \\ h\boldsymbol{C}_{\mathrm{a}i} & \boldsymbol{I} \end{pmatrix}, \quad \boldsymbol{B}_{\mathrm{z}i}(h_{\mathrm{act}_i}) = \begin{pmatrix} \boldsymbol{B}_{\mathrm{a}i}(h_{\mathrm{act}_i}) \\ 0 \end{pmatrix}, \quad \boldsymbol{C}_{\mathrm{z}i}(h_{\mathrm{act}_i}) = \begin{pmatrix} \boldsymbol{C}_{\mathrm{a}i} & 0 \\ 0 & \boldsymbol{I} \end{pmatrix},$$

where $\boldsymbol{A}_{\mathrm{a}i}(h_{\mathrm{act}_i})$, $\boldsymbol{B}_{\mathrm{a}i}(h_{\mathrm{act}_i})$ and $\boldsymbol{C}_{\mathrm{a}i}$ are defined according to (4.4). Under forward integration, the discrete-time weighting matrix $\boldsymbol{Q}_i$ is derived in an analog way as described in Appendix A.1.3. The control parameters of one control task T_i can then be determined applying the following theorem.

Theorem A.1 *The control parameters $\boldsymbol{K}_i = \left(\boldsymbol{K}_{\mathrm{P}i} \quad \boldsymbol{K}_{\mathrm{I}i}\right)$ are obtained from the LMI optimization problem*

$$\min_{\boldsymbol{Z}_i, \boldsymbol{K}_i} \operatorname{tr}\left(\boldsymbol{Z}_i^{-1}\right) \quad \text{subject to} \tag{A.18a}$$

$$\boldsymbol{V}_i\boldsymbol{C}_{\mathrm{z}i} = \boldsymbol{C}_{\mathrm{z}i}\boldsymbol{G}_i \tag{A.18b}$$

$$\begin{pmatrix} \boldsymbol{G}_i^T + \boldsymbol{G}_i - \boldsymbol{Z}_i & * & * \\ \boldsymbol{A}_{\mathrm{z}i}(h_{\mathrm{act}_i})\boldsymbol{G}_i + \boldsymbol{B}_{\mathrm{z}i}(h_{\mathrm{act}_i})\boldsymbol{U}_i\boldsymbol{C}_{\mathrm{z}i}(h_{\mathrm{act}_i}) & \boldsymbol{Z}_i & * \\ \boldsymbol{Q}_i^{1/2}\begin{pmatrix} \boldsymbol{G}_i \\ \boldsymbol{U}_i\boldsymbol{C}_{\mathrm{z}i} \end{pmatrix} & 0 & \boldsymbol{I} \end{pmatrix} > 0 \tag{A.18c}$$

with the LMI variables $\boldsymbol{G}_i \in \mathbb{R}^{(n_i+m_i+p_i)\times(n_i+m_i+p_i)}$, $\boldsymbol{V}_i \in \mathbb{R}^{2p_i\times 2p_i}$ and $\boldsymbol{U}_i \in \mathbb{R}^{m_i\times 2p_i}$ unrestricted, and $\boldsymbol{Z}_i = \boldsymbol{P}_i^{-1}$ symmetric positive definite. The control gain results from

$$\boldsymbol{K}_i = \boldsymbol{U}_i\boldsymbol{V}_i^{-1}. \tag{A.19}$$

PROOF. The proof follows the lines of the proof of Theorem 4.1 and Theorem 4.2. □

A.4 Control Design for the Related Methods of Event-Triggered PI Control

For comparative methods of event-triggered PI control, strategies for designing the control parameters are not given in literature. In order to be able to compare the results with the ones of the proposed event-triggered PI control approaches, the control parameters are designed minimizing an upper bound of the cost function (5.7), which is also termed *Guaranteed Cost PI Control* [XY04, MUTT09]. Thereby, a time-based implementation is assumed, i.e. for the simple ETC approach a conventional sampled-data control implementation is assumed for the design of the control parameters (Appendix A.4.1) and for the continuous-time ETC approach a continuous-time implementation is assumed (Appendix A.4.2). For the control design, the reference signal $\boldsymbol{r}$ is set zero.

A.4.1 Guaranteed Cost PI Control for Discrete-Time Systems

The aim is to find a control gain $\boldsymbol{K} = \begin{pmatrix}\boldsymbol{K}_\mathrm{P} & \boldsymbol{K}_\mathrm{I}\end{pmatrix}$ for the discrete-time system (5.6) with the PI control law (5.2) such that the closed loop system is GAS with guaranteed performance.

Theorem A.2 *The control parameters $\boldsymbol{K} = \begin{pmatrix}\boldsymbol{K}_\mathrm{P} & \boldsymbol{K}_\mathrm{I}\end{pmatrix}$ are obtained from the LMI optimization problem*

$$\min_{\boldsymbol{Z},\boldsymbol{K}} \operatorname{tr}\left(\boldsymbol{Z}^{-1}\right) \quad \text{subject to} \tag{A.20a}$$

$$\boldsymbol{V}\boldsymbol{C}_\mathrm{z} = \boldsymbol{C}_\mathrm{z}\boldsymbol{G} \tag{A.20b}$$

$$\begin{pmatrix} \boldsymbol{G}^T + \boldsymbol{G} - \boldsymbol{Z} & * & * \\ \boldsymbol{A}_\mathrm{z}\boldsymbol{G} + \boldsymbol{B}_\mathrm{z}\boldsymbol{U}\boldsymbol{C}_\mathrm{z} & \boldsymbol{Z} & * \\ \boldsymbol{Q}^{1/2}\begin{pmatrix}\boldsymbol{G} \\ \boldsymbol{U}\boldsymbol{C}_\mathrm{z}\end{pmatrix} & 0 & \boldsymbol{I} \end{pmatrix} > 0 \tag{A.20c}$$

with the LMI variables $\boldsymbol{G} \in \mathbb{R}^{(n+m+p)\times(n+m+p)}$, $\boldsymbol{V} \in \mathbb{R}^{2p\times 2p}$ and $\boldsymbol{U} \in \mathbb{R}^{m\times 2p}$ unrestricted, and $\boldsymbol{Z} = \boldsymbol{P}^{-1}$ symmetric positive definite. The control gain results from

$$\boldsymbol{K} = \boldsymbol{U}\boldsymbol{V}^{-1}. \tag{A.21}$$

Proof. The proof follows the lines of the proof of Theorem 5.1 and Theorem 5.2 in a simplified way. □

A.4.2 Guaranteed Cost PI Control for Continuous-Time Systems

For designing the control parameters of the continuous-time PI controller the input delay τ is neglected for simplicity. Controlling the continuous-time system (5.1) with the continuous-time PI controller

$$\dot{\boldsymbol{x}}_\mathrm{I}(t) = \boldsymbol{y}(t) - \boldsymbol{r} \tag{A.22a}$$

$$\boldsymbol{u}(t) = \boldsymbol{K}_\mathrm{I}\boldsymbol{x}_\mathrm{I}(t) + \boldsymbol{K}_\mathrm{P}\big(\boldsymbol{y}(t) - \boldsymbol{r}\big), \tag{A.22b}$$

an augmented system model can be built up, i.e.

$$\begin{pmatrix}\dot{\boldsymbol{x}}(t) \\ \dot{\boldsymbol{x}}_\mathrm{I}(t)\end{pmatrix} = \left[\begin{pmatrix}\boldsymbol{A}_\mathrm{p} & 0 \\ \boldsymbol{C}_\mathrm{p} & 0\end{pmatrix} + \begin{pmatrix}\boldsymbol{B}_\mathrm{p} \\ 0\end{pmatrix}\boldsymbol{K}\begin{pmatrix}\boldsymbol{C}_\mathrm{p} & 0 \\ 0 & \boldsymbol{I}\end{pmatrix}\right]\begin{pmatrix}\boldsymbol{x}(t) \\ \boldsymbol{x}_\mathrm{I}(t)\end{pmatrix} + \begin{pmatrix}-\boldsymbol{B}_\mathrm{p}\boldsymbol{K}_\mathrm{P} \\ -\boldsymbol{I}\end{pmatrix}\boldsymbol{r} \tag{A.23}$$

where $\boldsymbol{r} = \boldsymbol{0}$ for designing the control parameters. The aim is to find the control gain $\boldsymbol{K} = \begin{pmatrix}\boldsymbol{K}_\mathrm{P} & \boldsymbol{K}_\mathrm{I}\end{pmatrix}$ for the control law (A.22) such that the cost function

$$J = \int_0^\infty \begin{pmatrix}\boldsymbol{x}_\mathrm{p}(t) \\ \boldsymbol{x}_\mathrm{I}(t)\end{pmatrix}^T \begin{pmatrix}\boldsymbol{Q}_\mathrm{c} & 0 \\ 0 & \boldsymbol{Q}_\mathrm{I}\end{pmatrix} \begin{pmatrix}\boldsymbol{x}_\mathrm{p}(t) \\ \boldsymbol{x}_\mathrm{I}(t)\end{pmatrix} + \boldsymbol{u}^T(t)\boldsymbol{R}_\mathrm{c}\boldsymbol{u}(t)dt \tag{A.24}$$

is minimized. For first order systems, i.e. $\boldsymbol{C}_{\mathrm{p}} = \boldsymbol{I}$, the standard LQR method can be applied, see e.g. [KS72, Chapter 3]. For higher order systems, if $\boldsymbol{C}_{\mathrm{p}} \neq \boldsymbol{I}$, the algorithm given in [LVS12, Section 8.1] can be applied for determining the output feedback gain $\boldsymbol{K}$.

Bibliography

[ÅB02] K. J. Åström and B. Bernhardsson. Comparison of Riemann and Lebesgue sampling for first order stochastic systems. In *Proceedings of the 41th IEEE Conference on Decision and Control*, pages 2011–2016, 2002.

[ÅC05] K.-E. Årzén and A. Cervin. Control and embedded computing: Survey of research directions. In *Proceedings of 16th IFAC World Congress*, 2005.

[ÅCH03] K.-E. Årzén, A. Cervin, and D. Henriksson. Resource-constrained embedded control systems: Possibilities and research issues. In *Proceedings of CERTS'03 – Co-design of Embedded Real-Time Systems Workshop*, 2003.

[ACL05] K. H. Ang, G. Chong, and Y. Li. PID control system analysis, design, and technology. *IEEE Transactions on Control Systems Technology*, 13(4):559–576, 2005.

[AGL15] S. Al-Areqi, D. Görges, and S. Liu. Event-based networked control and scheduling codesign with guaranteed performance. *Automatica*, to appear, 2015.

[AGRL13] S. Al-Areqi, D. Görges, S. Reimann, and S. Liu. Event-based control and scheduling codesign of networked embedded control systems. In *Proceedings of the 2013 American Control Conference*, pages 5319–5324, 2013.

[Årz99] K.-E. Årzén. A simple event-based PID controller. In *Proceedings of 14th IFAC World Congress*, 1999.

[ÅW90] K. J. Åström and B. Wittenmark. *Computer-Controlled Systems: Theory and Design*. Prentice-Hall, Englewood Cliffs, NJ, 2nd edition, 1990.

[BC08] E. Bini and A. Cervin. Delay-aware period assignment in control systems. In *Proceedings of the 29th IEEE Real-Time Systems Symposium*, pages 291–300, 2008.

[BCD90] S. Bittanti, P. Colaneri, and G. De Nicolao. An algebraic Riccati equation for the discrete-time periodic prediction problem. *Systems & Control Letters*, 14(1):71–78, 1990.

[BCD91] S. Bittanti, P. Colaneri, and G. De Nicolao. The periodic Riccati equation. In S. Bittani, A. J. Laub, and J. D. Willems, editors, *The Riccati Equation*, Communications and Control Engineering, chapter 6, pages 127–162.

Springer, Berlin, 1991.

[BÇH06] M.-M. Ben Gaid, A. Çela, and Y. Hamam. Optimal integrated control and scheduling of networked control systems with communication constraints: Application to a car suspension system. *IEEE Transactions on Control Systems Technology*, 14(4):776–787, 2006.

[BÇH09] M.-M. Ben Gaid, A. Çela, and Y. Hamam. Optimal real-time scheduling of control tasks with state feedback resource allocation. *IEEE Transactions on Control Systems Technology*, 17(2):309–326, 2009.

[BEFB94] S. Boyd, L. El Ghaoui, E. Feron, and V. Balakrishnan. *Linear Matrix Inequalities in System and Control Theory*. Society for Industrial and Applied Mathematics (SIAM), Philadelphia, 1994.

[Ben06] M.-M. Ben Gaid. *Optimal Scheduling and Control for Distributed Real-Time Systems*. PhD thesis, Université d'Evry Val d'Essonne, Paris, France, 2006.

[BHJ10] A. Bemporad, W. P. M. H. Heemels, and M. Johansson. *Networked Control Systems*. Lecture Notes in Control and Information Sciences. Springer, London, 2010.

[Bla99] F. Blanchini. Set invariance in control. *Automatica*, 35(11):1747–1767, 1999.

[BMM09] F. Blanchini, S. Miani, and F. Mesquine. A separation principle for linear switching systems and parametrization of all stabilizing controllers. *IEEE Transactions on Automatic Control*, 54(2):279–292, 2009.

[Bra98] M. S. Branicky. Multiple Lyapunov functions and other analysis tools for switched and hybrid systems. *IEEE Transactions on Automatic Control*, 43(4):475–482, 1998.

[CA06] A. Cervin and P. Alriksson. Optimal on-line scheduling of multiple control tasks: A case study. In *Proceedings of the 18th Euromicro Conference on Real-Time Systems*, pages 141–150, 2006.

[CEBÅ02] A. Cervin, J. Eker, B. Bernhardsson, and K.-E. Årzén. Feedback-feedforward scheduling of control tasks. *Real-Time Systems*, 23(1-2):25–53, 2002.

[Cer03] A. Cervin. *Integrated Control and Real-Time Scheduling*. PhD thesis, Department of Automatic Control, Lund Institute of Technology, Lund, Sweden, 2003.

[CH08] A. Cervin and T. Henningsson. Scheduling of event-triggered controllers on a shared network. In *Proceedings of the 47th IEEE Conference on Decision and Control*, pages 3601–3606, 2008.

[CHL+03] A. Cervin, D. Henriksson, B. Lincoln, J. Eker, and K.-E. Årzén. How does control timing affect performance? *IEEE Control Systems Magazine*, 23(3):16–30, 2003.

[CMV+06] R. Castañé, P. Martí, M. Velasco, A. Cervin, and D. Henriksson. Resource management for control tasks based on the transient dynamics of closed-loop systems. In *Proceedings of the 18th Euromicro Conference on Real-Time Systems*, pages 171–182, 2006.

[Cog09] R. Cogill. Event-based control using quadratic approximate value functions. In *Proceedings of the 48th IEEE Conference on Decision and Control and 28th Chinese Control Conference*, pages 5883–5888, 2009.

[CVHN09] M. B. G. Cloosterman, N. Van de Wouw, W. P. M. H. Heemels, and H. Nijmeijer. Stability of networked control systems with uncertain time-varying delays. *IEEE Transactions on Automatic Control*, 54(7):1575–1580, 2009.

[CVMC11] A. Cervin, M. Velasco, P. Martí, and A. Camacho. Optimal online sampling period assignment: Theory and experiment. *IEEE Transactions on Control Systems Technology*, 19(4):902–910, 2011.

[DB05] R. C. Dorf and R. H. Bishop. *Modern control systems*. Pearson Prentice Hall, Upper Saddle River, NJ, 10th edition, 2005.

[DGD11] G. S. Deaecto, J. C. Geromel, and J. Daafouz. Dynamic output feedback $\mathcal{H}_\infty$ control of switched linear systems. *Automatica*, 47:1713–1720, 2011.

[DH12] M. Donkers and W. P. M. H. Heemels. Output-based event-triggered control with guaranteed $\mathcal{L}_\infty$-gain and improved decentralised event-triggering. *IEEE Transactions on Automatic Control*, 57(6):1362–1376, 2012.

[DHVH11] M. C. F. Donkers, W. P. M. H. Heemels, N. Van De Wouw, and L. Hetel. Stability analysis of networked control systems using a switched linear systems approach. *IEEE Transactions on Automatic Control*, 56(9):2101–2115, 2011.

[DM09] S. Durand and N. Marchand. Further results on event-based PID controller. In *Proceedings of the 2009 European Control Conference*, pages 1979–1984, 2009.

[DRI02] J. Daafouz, P. Riedinger, and C. Iung. Stability analysis and control synthesis for switched systems: a switched Lyapunov function approach. *IEEE Transactions on Automatic Control*, 47(11):1883–1887, 2002.

[DRI03] J. Daafouz, P. Riedinger, and C. Iung. Observer-based switched control design for discrete-time switched systems. In *Proceedings of European Control Conference*, 2003.

[EDK10] A. Eqtami, D. V. Dimarogonas, and K. J. Kyriakopoulos. Event-triggered control for discrete-time systems. In *Proceedings of the 2010 American Control Conference*, pages 4719–4724, 2010.

[EHÅ00] J. Eker, P. Hagander, and K.-E. Årzén. A feedback scheduler for real-time controller tasks. *Control Engineering Practice*, 8(12):1369–1378, 2000.

[ELLSV97] S. Edwards, L. Lavagno, E. A. Lee, and A. Sangiovanni-Vincentelli. Design of embedded systems: Formal models, validation, and synthesis. *Proceedings of the IEEE*, 85(3):366–390, 1997.

[FN01] L. M. Feeney and M. Nilsson. Investigating the energy consumption of a wireless network interface in an ad hoc networking environment. In *IEEE INFOCOM 2001. Proceedings of the Twentieth Annual Joint Conference of the IEEE Computer and Communications Societies*, pages 1548–1557, 2001.

[FPW98] G. F. Franklin, J. D. Powell, and M. Workman. *Digital control of dynamic systems*. Addison-Wesley, Reading, MA, 3rd edition, 1998.

[FTCMM02] G. Ferrari-Trecate, F. A. Cuzzola, D. Mignone, and M. Morari. Analysis of discrete-time piecewise affine and hybrid systems. *Automatica*, 38(12):2139–2146, 2002.

[GB08] M. Grant and S. Boyd. Graph implementations for nonsmooth convex programs. In V. Blondel, S. Boyd, and H. Kimura, editors, *Recent Advances in Learning and Control*, Lecture Notes in Control and Information Sciences, pages 95–110. Springer-Verlag Limited, 2008.

[GB12] M. Grant and S. Boyd. CVX: Matlab software for disciplined convex programming, version 2.0. `http://cvxr.com/cvx`, 2012.

[GC06] J. C. Geromel and P. Colaneri. Stability and stabilization of discrete time switched systems. *International Journal of Control*, 79(7):719–728, 2006.

[GCB08] J. C. Geromel, P. Colaneri, and P. Bolzern. Dynamic output feedback control of switched linear systems. *IEEE Transactions on Automatic Control*, 53(3):720–733, 2008.

[GIL07] D. Görges, M. Izák, and S. Liu. Optimal control of systems with resource constraints. In *Proceedings of the 46th IEEE Conference on Decision and Control*, pages 1070–1075, 2007.

[GIL09] D. Görges, M. Izák, and S. Liu. Optimal control and scheduling of networked control systems. In *Proceedings of the 48th IEEE Conference on Decision and Control and 28th Chinese Control Conference*, pages 5839–5844, 2009.

[GIL11] D. Görges, M. Izák, and S. Liu. Optimal control and scheduling of switched systems. *IEEE Transactions on Automatic Control*, 56(1):135–140, 2011.

[GLS14] J. M. Gomes da Silva Jr., W. F. Lages, and D. Sbarbaro. Event-triggered PI control design. In *Proceedings of 19th IFAC World Congress*, pages 6947–6952, 2014.

[Gör12] D. Görges. *Optimal Control of Switched Systems*. PhD thesis, Institute of Control Systems, Department of Electrical and Computer Engineering, Kaiserslautern, Germany, 2012.

[HC05] D. Henriksson and A. Cervin. Optimal on-line sampling period assignment for real-time control tasks based on plant state information. In *Proceedings of the 44th IEEE Conference on Decision and Control and European Control Conference*, pages 4469–4474, 2005.

[HD13] W. P. M. H. Heemels and M. C. F. Donkers. Model-based periodic event-triggered control for linear systems. *Automatica*, 49(3):698–711, 2013.

[HDT13] W. P. M. H. Heemels, M. C. R. Donkers, and A. R. Teel. Periodic event-triggered control for linear systems. *IEEE Transactions on Automatic Control*, 58(4):847–861, 2013.

[HNX07] J. P. Hespanha, P. Naghshtabrizi, and Y. Xu. A survey of recent results in networked control systems. *Proceedings of the IEEE*, 95(1):138–162, 2007.

[HPVK12] B. Hensel, J. Ploennigs, V. Vasyutynskyy, and K. Kabitzsch. A simple PI controller tuning rule for sensor energy efficiency with level-crossing sampling. In *9th International Multi-Conference on Systems, Signals and Devices (SSD)*, pages 1–6, 2012.

[JWX05] Z. Ji, L. Wang, and G. Xie. Quadratic stabilization of uncertain discrete-time switched systems via output feedback. *Circuits, Systems and Signal Processing*, 24(6):733–751, 2005.

[Kan97] M. Kantner. Robust stability of piecewise linear discrete-time systems. In *Proceedings of the 1997 American Control Conference*, pages 1241–1245, 1997.

[Kha02] H. K. Khalil. *Nonlinear Systems*, volume 3. Prentice-Hall, Upper Saddle River, New Jersey, 2002.

[KLJ14] G. A. Kiener, D. Lehmann, and K. H. Johansson. Actuator saturation and anti-windup compensation in event-triggered control. *Discrete event dynamic systems*, 24(2):173–197, 2014.

[KS72] H. Kwakernaak and R. Sivan. *Linear optimal control systems*. Wiley-interscience, New York, 1972.

[LA06] H. Lin and P. J. Antsaklis. Switching stabilization and l_2 gain performance controller synthesis for discrete-time switched linear systems. In *Proceeding of the 45th IEEE Conference on Decision and Control*, pages 2673–2678, 2006.

[LA09] H. Lin and P. J. Antsaklis. Stability and stabilizability of switched linear systems: A survey on recent results. *IEEE Transactions on Automatic Control*, 54(2):308–322, 2009.

[LB02] B. Lincoln and B. Bernhardsson. LQR optimization of linear system switching. *IEEE Transactions on Automatic Control*, 47(10):1701–1705, 2002.

[LJ12] D. Lehmann and K. H. Johansson. Event-triggered PI control subject to actuator saturation. In *IFAC Conference on Advances in PID Control*, 2012.

[LKJ12] D. Lehmann, G. A. Kiener, and K. H. Johansson. Event-triggered PI control: Saturating actuators and anti-windup compensation. In *Proceedings of the 51st IEEE Conference on Decision and Control*, pages 6566–6571, 2012.

[LL73] C. L. Liu and J. W. Layland. Scheduling algorithms for multiprogramming in a hard-real-time environment. *Journal of the ACM*, 20(1):46–61, 1973.

[LL10] J. Lunze and D. Lehmann. A state-feedback approach to event-based control. *Automatica*, 46(1):211–215, 2010.

[LL11a] D. Lehmann and J. Lunze. Event-based output-feedback control. In *19th Mediterranean Conference on Control and Automation*, pages 982–987, 2011.

[LL11b] D. Lehmann and J. Lunze. Extension and experimental evaluation of an event-based state-feedback approach. *Control Engineering Practice*, 19(2):101–112, 2011.

[LMT01] F.-L. Lian, J. R. Moyne, and D. M. Tilbury. Performance evaluation of control networks: Ethernet, controlnet, and devicenet. *IEEE Control Systems Magazine*, 21(1):66–83, 2001.

[LMV+13] C. Lozoya, P. Martí, M. Velasco, J. M. Fuertes, and E. X. Martin. Resource and performance trade-offs in real-time embedded control systems. *Real-Time Systems*, 49(3):267–307, 2013.

[LMVF12] C. Lozoya, P. Martí, M. Velasco, and J. M. Fuertes. Performance evaluation framework (PEF) for real-time embedded control systems. Research Report ESAII-RR-12-02, Automatic Control Department, Technical University of Catalonia, 2012.

[Löf04] J. Löfberg. YALMIP : a toolbox for modeling and optimization in MAT-

LAB. In *Proceedings of the 2004 IEEE International Symposium on Computer Aided Control Systems Design*, pages 284–289, 2004.

[LVS12] F. L. Lewis, D. Vrabie, and V. L. Syrmos. *Optimal control.* Hoboken, NJ: John Wiley & Sons, 3rd edition, 2012.

[MFTM00] D. Mignone, G. Ferrari-Trecate, and M. Morari. Stability and stabilization of piecewise affine and hybrid systems: an LMI approach. In *Proceedings of the 39th IEEE Conference on Decision and Control*, pages 504–509, 2000.

[MH09] A. Molin and S. Hirche. On LQG joint optimal scheduling and control under communication constraints. In *Proceedings of the 48th IEEE Conference on Decision and Control and 28th Chinese Control Conference*, pages 5832–5838, 2009.

[MH11] A. Molin and S. Hirche. Optimal design of decentralized event-triggered controllers for large-scale systems with contention-based communication. In *Proceedings of the 5th IEEE Conference on Decision and Control*, pages 4710–4716, 2011.

[MH14] A. Molin and S. Hirche. A bi-level approach for the design of event-triggered control systems over a shared network. *Discrete event dynamic systems*, 24(2):173–197, 2014.

[MLB+04] P. Martí, C. Lin, S. A. Brandt, M. Velasco, and J. M. Fuertes. Optimal state feedback based resource allocation for resource-constrained control tasks. In *Proceedings of the 25th IEEE International Real-Time Systems Symposium*, pages 161–172, 2004.

[MLB+09] P. Martí, C. Lin, S. A. Brandt, M. Velasco, and J. M. Fuertes. Draco: Efficient resource management for resource-constrained control tasks. *IEEE Transactions on Computers*, 58(1):90–105, 2009.

[MT11] M. Mazo and P. Tabuada. Dezentralized event-triggered control over wireless sensor/actuator networks. *IEEE Transactions on Automatic Control*, 56(10):2456–2461, 2011.

[MUTT09] H. Mukaidani, T. Umeda, Y. Tanaka, and T. Tsuji. Guaranteed cost PI control for uncertain discrete-time systems with additive gain. In *Proceedings of the 2009 European Control Conference*, pages 2319–2324, 2009.

[NFZE07] G. N. Nair, F. Fagnani, S. Zampieri, and R. J. Evans. Feedback control under data rate constraints: An overview. *Proceedings of the IEEE*, 95(1):108–137, 2007.

[OHC07] M. Ohlin, D. Henriksson, and A. Cervin. TrueTime 1.5 — reference manual. Department of Automatic Control, Lund University, 2007.

[OMT02] P. G. Otanez, J. G. Moyne, and D. M. Tilbury. Using deadbands to reduce communication in networked control systems. In *Proceedings of the 2002 American Control Conference*, pages 3015–3020, 2002.

[Pet03] S. Pettersson. Synthesis of switched linear systems. In *Proceedings of the 42nd IEEE Conference on Decision and Control*, pages 5283–5288, 2003.

[RAL13a] S. Reimann, S. Al-Areqi, and S. Liu. An event-based online scheduling approach for networked embedded control systems. In *Proceedings of the 2013 American Control Conference*, pages 5346–5351, 2013.

[RAL13b] S. Reimann, S. Al-Areqi, and S. Liu. Output-based control and scheduling codesign for control systems sharing a limited resource. In *Proceedings of the 52nd IEEE Conference on Decision and Control*, pages 4042–4047, 2013.

[RJ08] M. Rabi and K. H. Johansson. Event-triggered strategies for industrial control over wireless networks. In *International conference on wireless internet*, pages 1–7, 2008.

[RJSD14] Á. Ruiz, J. E. Jiménez, J. Sánchez, and S. Dormido. Design of event-based PI-P controllers using interactive tools. *Control Engineering Practice*, 32:183–202, 2014.

[RM09] J. B. Rawlings and D. Q. Mayne. *Model Predictive Control: Theory and Design*. Nob Hill Publishing, Madison, WI, 2009.

[RS00] H. Rehbinder and M. Sanfridson. Integration of off-line scheduling and optimal control. In *Proceedings of the 12th Euromicro Conference on Real-Time Systems*, pages 137–143, 2000.

[RS04] H. Rehbinder and M. Sanfridson. Scheduling of a limited communication channel for optimal control. *Automatica*, 40(3):491–500, 2004.

[RV07] D. Rosinová and V. Veselý. Robust PID decentralized controller design using LMI. *International Journal of Computers, Communications & Control*, 2(2):195–204, 2007.

[RVAL15] S. Reimann, D. H. Van, S. Al-Areqi, and S. Liu. Stability analysis and PI control synthesis under event-triggered communication. In *Proceedings of the 2015 European Control Conference*, 2015.

[RWL12] S. Reimann, W. Wu, and S. Liu. A novel control-schedule codesign method for embedded control systems. In *Proceedings of the 2012 American Control Conference*, pages 3766–3771, 2012.

[RWL14a] S. Reimann, W. Wu, and S. Liu. PI control and scheduling design for embedded control systems. In *Proceedings of 19th IFAC World Congress*, pages 11111–11116, 2014.

[RWL14b] S. Reimann, W. Wu, and S. Liu. PI control synthesis for event-triggered control systems. *Submitted for journal publication*, 2014.

[RWL14c] S. Reimann, W. Wu, and S. Liu. Real-time scheduling of PI control tasks. *Submitted for journal publication*, 2014.

[San04] M. Sanfridson. *Quality of control and real-time scheduling*. PhD thesis, Royal Institute of Technology, Department of Machine Design, Stockholm, Sweden, 2004.

[SBR05] A. Suri, J. Baillieul, and D. V. Raghunathan. Control using feedback over wireless ethernet and bluetooth. In D. Hristu-Varsakelis and W. S. Levine, editors, *Handbook of Networked and Embedded Control Systems*, pages 677–697. Birkhäuser, Boston, MA, 2005.

[SK97] M. Stemm and R. H. Katz. Measuring and reducing energy consumption of network interfaces in hand-held devices. *IEICE Transactions on Communications*, 80(8):1125–1131, 1997.

[SLSS96] D. Seto, J. P. Lehoczky, L. Sha, and K. G. Shin. On task schedulability in real-time control systems. In *17th IEEE Real-Time Systems Symposium*, pages 13–21, 1996.

[SP05] S. Skogestad and I. Postlethwaite. *Multivariable Feedback Control: Theory and Design*. Wiley, Chichester, 2nd edition, 2005.

[SPTZ13] A. Seuret, C. Prieur, S. Tarbouriech, and L. Zaccarian. Event-triggered control with LQ optimality guarantees for saturated linear systems. In *IFAC Symposium on Nonlinear Control*, 2013.

[SSA10] D. Simon, Y.-Q. Song, and C. Auburn, editors. *Co-design Approaches to Dependable Networked Control Systems*. ISTE and Wiley, 2010.

[Stu99] J. F. Sturm. Using SeDuMi 1.02, a MATLAB toolbox for optimization over symmetric cones (updated for version 1.34). *Optimization Methods and Software*, 11(1-4):625–653, 1999.

[SVD11] J. Sánchez, A. Visioli, and S. Dormido. A two-degree-of-freedom PI controller based on events. *Journal of Process Control*, 21(4):639–651, 2011.

[Tab07] P. Tabuada. Event-triggered real-time scheduling of stabilizing control tasks. *IEEE Transactions on Automatic Control*, 52(9):1680–1685, 2007.

[TAJ12] U. Tiberi, J. Araújo, and K. H. Johansson. On event-based PI control of first-order processes. In *IFAC Conference on Advances in PID Control*, 2012.

[TGGQ11] S. Tarbouriech, G. Garcia, J. M. Gomes da Silva Jr., and I. Queinnec. *Stability and Stabilization of Linear Systems with Saturating Actuators*.

Springer, 2011.

[TPF97] R. H. C. Takahashi, P. L. D. Peres, and P. A. V. Ferreira. Multiobjective H_2/H_∞ guaranteed cost PID design. *IEEE Control Systems*, 17(5):37–47, 1997.

[VK06] V. Vasyutynskyy and K. Kabitzsch. Implementation of PID controller with send-on-delta sampling. In *Proceedings of International control conference*, Glasgow, 2006.

[WL11] X. Wang and M. D. Lemmon. Event-triggering in distributed networked control systems. *IEEE Transactions on Automatic Control*, 56(3):586–601, 2011.

[WRGL15] W. Wu, S. Reimann, D. Görges, and S. Liu. Suboptimal event-triggered control for time-delayed linear systems. *IEEE Transactions on Automatic Control*, to appear, 2015.

[WRL14] W. Wu, S. Reimann, and S. Liu. Event-triggered control for linear systems subject to actuator saturation. In *Proceedings of 19th IFAC World Congress*, pages 9492–9497, 2014.

[XY04] J.-M. Xu and L. Yu. An LMI approach to guaranteed cost PI control of linear uncertain systems. In *Proceedings of the 43rd IEEE Conference on Decision and Control*, pages 2165–2170, 2004.

[ZAHV09] W. Zhang, A. Abate, J. Hu, and M. P. Vitus. Exponential stabilization of discrete-time switched linear systems. *Automatica*, 45(11):2526–2536, 2009.

[Zha01] G. Zhai. Quadratic stabilizability of discrete-time switched systems via state and output feedback. In *Proceeding of the 40th IEEE Conference on Decision and Control*, pages 2165–2166, 2001.

Zusammenfassung

Ausgangsbasierte Regelungs- und Schedulingverfahren für Prozesse mit Ressourcenbeschränkungen

Moderne Regelungen werden heutzutage meist in eingebetteten Systemen realisiert, wodurch eine hohe Funktionalität und eine flexible Anwendbarkeit erzielt wird. Ein eingebettetes System bezeichnet einen Rechner, der in einen physikalischen Kontext eingebunden ist. Anwendungen von eingebetteten Systemen finden sich in unterschiedlichen technischen Anwendungen, unter anderem Verkehrsmittel wie Kraftfahrzeuge und Flugzeuge, Unterhaltungselektronik wie CD-Spieler, industrielle Anlagen wie Fertigungsanlagen und verfahrenstechnische Anlagen, sowie Infrastruktursysteme wie Energienetze.

In eingebetteten Regelungssystemen arbeitet gewöhnlich ein Rechner mehrere Regelungsprozesse sowie weitere Prozesse ab. Dies macht ein sogenanntes Scheduling erforderlich, welches den Prozessen die zeitlich begrenzten Rechenressourcen zuteilt. Im Weiteren treten moderne Regelungen zunehmend mit einer verteilten Struktur auf. Dabei sind Sensoren, Aktoren und Regler räumlich verteilt und kommunizieren über ein Netzwerk. Die Vernetzung von Regelungssystemen bringt Vorteile wie erhöhte Flexibilität und Instandhaltbarkeit mit sich. Auf der anderen Seite ist eine Medium Access Control erforderlich, welche den Zugriff auf die Kommunikationsressource steuert.

Charakteristisch für eingebettete Regelungssysteme ist neben der Flexibilität die Beschränktheit der Ressourcen, wie Rechen- und Kommunikationsressourcen. Dafür gibt es mehrere Gründe. Zum einen werden eingebettete Systeme insbesondere in Massenprodukten eingesetzt. Durch den starken Wettbewerb auf dem Markt, spielen in der Entwicklung insbesondere die Kosten der einzelnen Komponenten eine Rolle, was begrenzte Rechenleistung, Speichergröße und Kommunikationsbandbreite mit sich bringt. Weitere Einschränkungen können sich durch die spezifische Anwendung ergeben. Für batteriebetriebene mobile Systeme, welche über ein drahtloses Netzwerk kommunizieren, ist der Energieverbrauch von Bedeutung. Dieser kann begrenzt werden, durch eine Reduktion der Kommunikation. Diese Arbeit behandelt Konzepte, um die begrenzten Ressourcen effizient für die Regelung zu nutzen. Dafür werden intelligente Schedulingverfahren sowie die ereignisbasierte Regelung vorgestellt.

Die Idee der ereignisbasierten Regelung ist es, die Stellgröße nicht periodisch zu aktualisieren, sondern nur dann wenn die Erforderlichkeit aus regelungstechnischer Sicht besteht. Dies wird durch eine Ereignisbedingung festgelegt, welche so definiert wird, dass der Kommunikationsaufwand beispielsweise deutlich reduziert werden kann, während die Regelgüte nahezu erhalten bleibt.

Diese Dissertation hat das Ziel, einen Beitrag zu Schedulingverfahren sowie zur ereignisbasierten Regelung zu leisten. Beide Ansätze erlauben einen effizienten Umgang mit den begrenzten Ressourcen und können gleichermaßen eine bestimmte Regelgüte garantieren. Der erste Teil der Arbeit (Kapitel 2-4) beschäftigt sich mit Schedulingverfahren, während der zweite Teil (Kapitel 5) auf die ereignisbasierte Regelung fokussiert.

In Kapitel 2 wird die Regelungsarchitektur beschrieben, in der mehrere Regelstrecken durch einen Prozessor geregelt werden. Dabei wird detailliert auf die einzelnen Schritte der Regelung und des Schedulings eingegangen und der Ressourcenaufwand aufgezeigt. Die Architektur wird anschließend noch durch ein Kommunikationsnetzwerk erweitert, über welches die Stellgrößensignale vom Prozessor zu den Aktoren gesendet werden.

Kapitel 3 ist fokussiert auf die Entwicklung eines integrierten Regler-Scheduler Entwurfs. Durch das integrierte Verfahren werden die Regler sowie der Netzwerkzugriff gemeinsam entworfen. Der Netzwerkzugriff wird dabei durch ein Online-Scheduling realisiert, d.h. der Zugriff wird basierend auf Informationen der Strecken online festgelegt. Sowohl die Regelung als auch das Scheduling soll durch Messung der Ausgangsgrößen der Strecken realisiert werden. Dafür wird in einem ersten Schritt ein Model des Gesamtsystems, bestehend aus den zu regelnden Strecken, dem Prozessor und dem Netzwerk, entwickelt. Nach der Diskretisierung der Zustandsraumgleichungen der einzelnen Strecken kann das Gesamtsystem als geschaltetes lineares zeitdiskretes System modelliert werden. Basierend darauf, wird ein Verfahren entwickelt, um Zustandsregler und einen Scheduler gemeinsam zu entwerfen. Dafür wird ein linear quadratisches Kostenfunktional als Gütekriterium eingeführt. Anschließend wird noch ein geschalteter Beobachter entworfen, um die erforderlichen Zustände zu schätzen. Da es sich um ein geschaltetes System handelt, muss noch nachgewiesen werden, dass das Separationsprinzip dafür gültig ist. Im Weiteren wird noch gezeigt, dass wenn die Zustandsregler und der Scheduler basierend auf den geschätzten Zuständen ausgeführt wird, eine bestimmte Regelgüte garantiert werden kann.

Abschließend wird die Methodik anhand von Simulationen und praktischen Implementierungen evaluiert. In der Simulation werden drei inverse Pendel über ein gemeinsames Netzwerk geregelt. Es werden impulsive Störungen auf die Strecken gegeben, auf die das vorgestellte Online-Scheduling besser reagiert als ein Offline-Scheduling und somit eine bessere Regelgüte aufweist. In der praktischen Implementierung werden zum einen ein inverses Pendel und zwei Gleichstrommotoren als Strecken betrachtet. Zum anderen werden zwei elektrische Doppelintegratorschaltungen betrachtet.

In Kapitel 4 stehen die begrenzten Rechenressourcen im Vordergrund. Während der in Kapitel 3 entworfene Scheduler einen hohen Rechenaufwand mit sich bringt und daher insbesondere bei vernetzten Regelungssystemen von Nutzen ist, wird in Kapitel 4 ein recheneffizienter Online-Scheduler entworfen. Dabei werden die zu regelnden Strecken jeweils durch einen PI-Regler geregelt. Dafür wird zuerst die diskrete Implementierung des PI Reglers adaptiert, so dass basierend darauf ein Online-Scheduler entworfen werden kann. Dieser verwendet die gemessenen Ausgangsgrößen, um den auszuführenden Regler zu bestimmen. Die Messung der kompletten Systemzustände ist nicht erforderlich. Anschließend wird ein Verfahren zur Bestimmung der Regelparameter vorgestellt, welches auf einem linear quadratischen Kostenfunktional basiert.

Das Verfahren wird abschließend durch eine Simulation sowie eine praktische Implementierung evaluiert. In der Simulation werden vier Strecken durch einen Prozessor geregelt. Vergleiche werden mit einem Offline-Scheduling Verfahren und mit dem EDF-Scheduling Verfahren durchgeführt. Eine praktische Evaluierung wird durch die Regelung der Drehzahl von drei Gleichstrommotoren mit einem Mikrocontroller realisiert.

Kapitel 5 beschäftigt sich mit der ereignisbasierten Regelung für PI-Regler mit dem Ziel, einen konstanten Sollwert einzuregeln. Dabei wird ebenfalls angenommen, dass nur die Ausgangsgrößen messbar sind. In der ereignisbasierten PI-Regelung treten mehrere Problemstellungen auf. In der Mehrheit der vorhandenen Ansätze ist die kontinuierliche Überprüfung der Ereignisbedingung erforderlich, was in einer praktischen Realisierung kritisch ist. Außerdem kann die ereignisbasierte PI-Regelung häufig zu einer Oszillation um den Sollwert oder zu einer stationären Abweichung vom Sollwert führen. Der Beitrag dieses Kapitels ist es, ein Verfahren vorzustellen, welches diese Probleme löst. Dafür wird zuerst die Struktur der ereignisbasierten PI-Regelung vorgestellt. Anschließend werden zwei Ereignisbedingungen definiert. Basierend darauf wird eine Reglersynthese entworfen, die eine effiziente Bestimmung der Regelparameter erlaubt. Dabei wird die ereignisbasierte Implementierung der PI-Regler berücksichtigt. Die Reglersynthese wird als Optimierungsproblem formuliert, wofür ein linear quadratisches Kostenfunktional als Zielfunktion eingeführt wird. Anschließend wird ein Stabilitätskriterium angegeben, anhand dessen die Stabilität analysiert werden kann.

Zur Evaluierung wird ein Vergleich mit zwei Methoden aus der Literatur anhand von Simulationen und praktischen Experimenten durchgeführt. Der Vergleich erfolgt basierend auf zwei Kriterien, der Regelgüte und der Anzahl der ausgelösten Ereignisse. Die Anzahl der ausgelösten Ereignisse repräsentiert den Kommunikationsaufwand, welcher durch die ereignisbasierte Regelung reduziert werden kann. In der Simulation wird die ereignisbasierte PI-Regelung an einem instabilen System erster Ordnung und einem stabilen System zweiter Ordnung angewandt. Die in der Dissertation eingeführte Methodik weist einen geringeren Kommunikationsaufwand und eine leicht bessere Regelgüte als die zum Vergleich herangezogenen Ansätze auf. In der praktischen Evaluierung zeigen sich ähnliche Ergebnisse. Dabei wird die ereignisbasierte PI-Regelung zur Drehzahlregelung eines Gleichstrommotors auf einem Mikrocontroller implementiert.

Kapitel 6 fasst die Arbeit zusammen und gibt einen Ausblick für künftige Forschungsrichtungen.

Curriculum Vitae

Education

10/2003 – 09/2009	**University of Kaiserslautern**
	Subject: Wirtschaftingenieurwesen (Business Administration and Engineering) Specialization: Automatisierungstechnik (Automation)
	Degree: Dipl.-Wirtsch.-Ing.
	Thesis: Regelung eines mobilen Manipulators zur Interaktion mit der Umgebung (Preis der Stiftung Pfalzmetall 2010)
09/1993 – 06/2002	**Gymnasium Herzogenaurach**
	Degree: Abitur
09/1989 – 07/1993	**Grundschule Herzogenaurach**

Professional Experience

10/2009 – 02/2015	**University of Kaiserslautern**
	Department of Electrical and Computer Engineering Institute of Control Systems
	Research Associate
04/2003 – 08/2003	**Siemens AG**
	Internship (Mechanical Manufacturing Methods)

In der Reihe *„Forschungsberichte aus dem Lehrstuhl für Regelungssysteme“*, herausgegeben von Steven Liu, sind bisher erschienen:

1	Daniel Zirkel	Flachheitsbasierter Entwurf von Mehrgrößenregelungen am Beispiel eines Brennstoffzellensystems ISBN 978-3-8325-2549-1, 2010, 159 S. 35.00 €
2	Martin Pieschel	Frequenzselektive Aktivfilterung von Stromoberschwingungen mit einer erweiterten modellbasierten Prädiktivregelung ISBN 978-3-8325-2765-5, 2010, 160 S. 35.00 €
3	Philipp Münch	Konzeption und Entwurf integrierter Regelungen für Modulare Multilevel Umrichter ISBN 978-3-8325-2903-1, 2011, 183 S. 44.00 €
4	Jens Kroneis	Model-based trajectory tracking control of a planar parallel robot with redundancies ISBN 978-3-8325-2919-2, 2011, 279 S. 39.50 €
5	Daniel Görges	Optimal Control of Switched Systems with Application to Networked Embedded Control Systems ISBN 978-3-8325-3096-9, 2012, 201 S. 36.50 €
6	Christoph Prothmann	Ein Beitrag zur Schädigungsmodellierung von Komponenten im Nutzfahrzeug zur proaktiven Wartung ISBN 978-3-8325-3212-3, 2012, 118 S. 33.50 €
7	Guido Flohr	A contribution to model-based fault diagnosis of electro-pneumatic shift actuators in commercial vehicles ISBN 978-3-8325-3338-0, 2013, 139 S. 34.00 €

8	Jianfei Wang	Thermal Modeling and Management of Multi-Core Processors ISBN 978-3-8325-3699-2, 2014, 144 S. 35.50 €
9	Stefan Simon	Objektorientierte Methoden zum automatisierten Entwurf von modellbasierten Diagnosesystemen ISBN 978-3-8325-3940-5, 2015, 197 S. 36.50 €
10	Sven Reimann	Output-Based Control and Scheduling of Resource-Constrained Processes ISBN 978-3-8325-3980-1, 2015, 145 S. 34.50 €